Problems in Molecular Orbital Theory

Problems in Molecular Orbital Theory

THOMAS A. ALBRIGHT
University of Houston

JEREMY K. BURDETT
The University of Chicago

New York   Oxford
OXFORD UNIVERSITY PRESS
1992

Oxford University Press

Oxford   New York   Toronto
Delhi   Bombay   Calcutta   Madras   Karachi
Kuala Lumpur   Singapore   Hong Kong   Tokyo
Nairobi   Dar es Salaam   Cape Town
Melbourne   Auckland   Madrid

and associated companies in
Berlin   Ibadan

Published by Oxford University Press, Inc.,
200 Madison Avenue, New York, New York 10016

Oxford is a registered trademark of Oxford University Press

Library of Congress Cataloging-in-Publication Data
Albright, Thomas A.
Problems in molecular orbital theory /
Thomas A. Albright, Jeremy K. Burdett.
p.   cm. Includes bibliographical references.
ISBN 0-19-507175-1
1. Molecular orbitals—Problems, exercises, etc. I. Burdett,
Jeremy K., 1947–   . II. Title.
QD461.A3843   1992   92-25944
541.2'24—dc20

1 3 5 7 9 8 6 4 2

Printed in the United States of America
on acid-free paper

PREFACE

The Cavendish Laboratory in Cambridge, England, has long published a small book called 'Cavendish Problems in Classical Physics'. It is designed to develop technique in a way, as its authors point out, like the five-finger exercises that are a vital step towards concert-hall mastery. The inclusion of problems at the end of chapters or sections in textbooks play a related rôle; here ensuring knowledge of the material by often repetetive drill. The set of problems in this book, from the area of Molecular Orbital Theory, though certainly narrower in scope, tries to provide similar opportunities.

One of the features of this book is that questions are collected from all areas of chemistry; inorganic, organic, transition metal, main group, organometallic, solid-state and molecular chemistry, static and dynamic problems. Although for convenience the questions are separated by area (albeit often quite tenuously), it is the authors' belief that an increased level of understanding can be achieved by removing the dividers which often separate one area of chemistry from another. It is particularly pleasing to see the unity of nature at work linking apparently unrelated observations together. For example, exactly the same orbital effect lies behind the planarity of trisilyamine, the anomeric effect of the organic chemist and the geometry around the oxygen atoms in the solid-state structure of the mineral rutile.

The Cavendish problems are divided into two types, those with an asterisk requiring a more mature appreciation of the principles behind the question. (A particularly elegant answer to one starred question is to be found in an early edition of The Encyclopædia Brittanica.) Some of the problems in this volume are of a similar type and are analogously identified with a *. A few of our problems here require numerical solution and are identified by a ‡. In all cases the mathematical problem is one of matrix diagonalization, for which there is ample software available to the reader with access to a personal computer. Some of the questions are quire similar and ask different, but related questions, about the same molecule, or lead the reader *via* a different pathway to the same eventual solution. We have had no hesitation in adapting an existing problem to our needs which we have found elsewhere. In fact some of the questions included here are quite 'old chestnuts', and are included for the obvious reasons. Although the division of the questions into chapters is done along traditional lines, for several questions the choice was somewhat arbitrary. We have made no attempt to organize the questions by subject matter or degree of difficulty within each chapter, and thus hope that the reader will sample the offerings as much as possible.

Obviously this book relies upon a basic knowledge of molecular orbital theory, discussed at various levels in many places[1-9]. Only in a few questions do we trudge through the labor of establishing the symmetry species appropriate for collections of orbitals. In an even smaller number of examples do we show the construction of symmetry adapted orbital sets. We therefore assume a certain basic knowledge of group theory, by and large, that of Cotton's book[5]. The exception is that symmetric and antisymmetric direct products are used to generate the symmetries of electronic states and to decide on the species of Jahn-Teller active vibrations. There is a brief discussion of such techniques in ref 1.

Most of the pictures in this book are hand-drawn, in an effort to show that clear molecular orbital diagrams and sketches of molecules *can* be drawn without the aid of an artist or computer graphics. In fact the sketching of the atomic orbital composition from the numerical results of a molecular orbital calculation (as in Question 4.16) leads to some valuable insights.

We thank the several echelons of students who have worked on these problems as well as Nita Yack and Natalie Silberman-Wainwright who typed the first draft of the manuscript. In addition we thank Roald Hoffmann and Timothy Hughbanks from whom we 'borrowed' some of the problems. Finally we thank I and J.

Chicago and Houston
July 1992

1. *Orbital Interactions in Chemistry*, T. A. Albright, J. K. Burdett and M.-H. Whangbo, Wiley (1985).
2. *Molecular Shapes*, J. K. Burdett, Wiley (1980).
3. *Electronic Spectra of Polyatomic Molecules*, G. Herzberg, Van Nostrand (1966).
4. *Molecular Structure and Bonding*, B. M. Gimarc, Academic Press (1979).
5. *Chemical Applications of Group Theory*, F. A. Cotton, Third Edition, Wiley (1990).
6. *The Chemical Bond*, J. N. Murrell, S. F. A. Kettle, J. M. Tedder, Second Edition, Wiley (1985).
7. *Hückel Molecular Orbital Theory*, K. Yates, Academic Press (1978).
8. *Molecular Electronic Structures*, C. J. Ballhausen, H. B. Gray, Benjamin-Cummings (1980).
9. *Frontier Orbitals and Organic Chemical Reactions*, I. Fleming, Wiley (1976).
10. *The HMO Model and its Application*, E. Heilbronner and H. Bock, Verlag Chemie (1976).

CONTENTS

Problems in Molecular
Orbital Theory

Chapter I.

Useful Supplementary Material

1.1 Selected Character Tables

C_1	E
A	1

C_s	E	σ_h		
A'	1	1	x, y, R_z	x^2, y^2, z^2, xy
A''	1	-1	z, R_x, R_y	xz, yz

C_i	E	i		
A_g	1	1	R_x, R_y, R_z	$x^2, y^2, z^2, xy, xz, yz$
A_u	1	-1	x, y, z	

C_2	E	C_2		
A	1	1	z, R_z	x^2, y^2, z^2, xy
B	1	-1	x, y, R_x, R_y	xz, yz

C_{2v}	E	C_2	$\sigma_v(xz)$	$\sigma_v(yz)$		
A_1	1	1	1	1	z	x^2, y^2, z^2
A_2	1	1	-1	-1	R_z	xy
B_1	1	-1	1	-1	x, R_y	xz
B_2	1	-1	-1	1	y, R_x	yz

C_{3v}	E	$2C_3$	$3\sigma_v$		
A_1	1	1	1	z	$x^2 + y^2, z^2$
A_2	1	1	-1	R_z	xy
E	2	-1	0	$(x, y)(R_x, R_y)$	$(x^2y^2, xy)(xz, yz)$

C_{4v}	E	$2C_4$	C_2	$3\sigma_v$	$3\sigma_d$		
A_1	1	1	1	1	1	z	$x^2 + y^2, z^2$
A_2	1	1	1	-1	-1	R_z	
B_1	1	-1	1	1	-1		$x^2 - y^2$
B_2	1	-1	1	-1	1		xy
E	2	0	-2	0	0	$(x, y)(R_x, R_y)$	(xz, yz)

C_{5v}	E	$2C_5$	$2C_5^2$	$5\sigma_v$		
A_1	1	1	1	1	z	$x^2 + y^2, z^2$
A_2	1	1	1	-1	R_z	
E_1	2	$2\cos72°$	$2\cos144°$	0	$(x, y)(R_x, R_y)$	(xz, yz)
E_2	2	$2\cos144°$	$2\cos72°$	0		$(x^2 - y^2, xy)$

C_{6v}	E	$2C_6$	$2C_3$	C_2	$3\sigma_v$	$3\sigma_d$		
A_1	1	1	1	1	1	1	z	$x^2 + y^2, z^2$
A_2	1	1	1	1	-1	-1	R_z	
B_1	1	-1	1	-1	1	-1		
B_2	1	-1	1	-1	-1	1		
E_1	2	1	-1	-2	0	0	$(x, y)(R_x, R_y)$	(xz, yz)
E_2	2	-1	-1	2	0	0		$(x^2 - y^2, xy)$

C_{2h}	E	C_2	i	σ_h		
A_g	1	1	1	1	R_z	x^2, y^2, z^2, xy
B_g	1	-1	1	-1	R_x, R_y	xz, yz
A_u	1	1	-1	-1	z	
B_u	1	-1	-1	1	x, y	

D_{2h}	E	$C_2(z)$	$C_2(y)$	$C_2(x)$	i	$\sigma_h(xy)$	$\sigma_h(xz)$	$\sigma_h(yz)$		
A_g	1	1	1	1	1	1	1	1		x^2, y^2, z^2
B_{1g}	1	1	-1	-1	1	1	-1	-1	R_z	xy
B_{2g}	1	-1	1	-1	1	-1	1	-1	R_y	xz
B_{3g}	1	-1	-1	1	1	-1	-1	1	R_x	yz
A_u	1	1	1	1	-1	-1	-1	-1		
B_{1u}	1	1	-1	-1	-1	-1	1	1	z	
B_{2u}	1	-1	1	-1	-1	1	-1	1	x	
B_{3u}	1	-1	-1	1	-1	1	1	-1	y	

D_{3h}	E	$2C_3$	$3C_2$	σ_h	$2S_3$	$3\sigma_v$		
A_1'	1	1	1	1	1	1		$x^2 + y^2, z^2$
A_2'	1	1	-1	1	1	-1	R_z	
E'	2	-1	0	2	-1	0	(x,y)	$(x^2 - y^2, xy)$
A_1''	1	1	1	-1	-1	-1		
A_2''	1	1	-1	-1	-1	1	z	
E''	2	-1	0	-2	1	0	(R_x, R_y)	(xz, yz)

D_{4h}	E	$2C_4$	C_2	$2C_2'$	$2C_2''$	i	$2S_4$	σ_h	$2\sigma_v$	$2\sigma_d$		
A_{1g}	1	1	1	1	1	1	1	1	1	1		$x^2 + y^2, z^2$
A_{2g}	1	1	1	-1	-1	1	1	1	-1	-1	R_z	
B_{1g}	1	-1	1	1	-1	1	-1	1	1	-1		$x^2 - y^2$
B_{2g}	1	-1	1	-1	1	1	-1	1	-1	1		xy
E_g	2	0	-2	0	0	2	0	-2	0	0	(R_x, R_y)	(xz, yz)
A_{1u}	1	1	1	1	1	-1	-1	-1	-1	-1		
A_{2u}	1	1	1	-1	-1	-1	-1	-1	1	1	z	
B_{1u}	1	-1	1	1	-1	-1	1	-1	-1	1		
B_{2u}	1	-1	1	-1	1	-1	1	-1	1	-1		
E_u	2	0	-2	0	0	-2	0	2	0	0	(x,y)	

D_{5h}	E	$2C_5$	$2C_5^2$	$5C_2$	σ_h	$2S_5$	$2S_5^2$	$5\sigma_v$		
A_1'	1	1	1	1	1	1	1	1		$x^2 + y^2, z^2$
A_2'	1	1	1	-1	1	1	1	-1	R_z	
E_1'	2	j	k	0	2	j	k	0	(x,y)	
E_2'	2	k	j	0	2	k	j	0		$(x^2 - y^2, xy)$
A_1''	1	1	1	1	-1	-1	-1	-1		
A_2''	1	1	1	-1	-1	-1	-1	1	z	
E_1''	2	j	k	0	-2	$-j$	$-k$	0	(R_x,R_y)	(xz,yz)
E_2''	2	k	j	0	-2	$-k$	$-j$	0		

$j = 2\cos72°, k = 2\cos144°$

D_{6h}	E	$2C_6$	$2C_3$	C_2	$3C_2'$	$3C_2''$	i	$2S_3$	$2S_6$	σ_h	$3\sigma_d$	$3\sigma_v$		
A_{1g}	1	1	1	1	1	1	1	1	1	1	1	1		$x^2 + y^2, z^2$
A_{2g}	1	1	1	1	-1	-1	1	1	1	1	-1	-1	R_z	
B_{1g}	1	-1	1	-1	1	-1	1	-1	1	-1	1	-1		
B_{2g}	1	-1	1	-1	-1	1	1	-1	1	-1	-1	1		
E_{1g}	2	1	-1	2	0	0	2	1	-1	2	0	0	(R_x,R_y)	(xz,yz)
E_{2g}	2	-1	-1	2	0	0	2	-1	-1	2	0	0		$(x^2 - y^2, xy)$
A_{1u}	1	1	1	1	1	1	-1	-1	-1	-1	-1	-1		
A_{2u}	1	1	1	1	-1	-1	-1	-1	-1	-1	1	1	z	
B_{1u}	1	-1	1	-1	1	-1	-1	1	-1	1	-1	1		
B_{2u}	1	-1	1	-1	-1	1	-1	1	-1	1	1	-1		
E_{1u}	2	1	-1	2	0	0	-2	-1	1	-2	0	0	(x,y)	
E_{2u}	2	-1	-1	2	0	0	-2	1	1	-2	0	0		

D_{2d}	E	$2S_4$	C_2	C_2'	$2\sigma_d$		
A_1	1	1	1	1	1		$x^2 + y^2, z^2$
A_2	1	1	1	-1	-1	R_z	
B_1	1	-1	1	1	-1		$(x^2 - y^2, xy)$
B_2	1	-1	1	-1	1	z	xy
E	2	0	-2	0	0	(x,y)	(xz,yz)
						(R_x,R_y)	

D_{3d}	E	$2C_3$	$3C_2$	i	$2S_6$	$3\sigma_d$		
A_{1g}	1	1	1	1	1	1		$x^2 + y^2, z^2$
A_{2g}	1	1	-1	1	1	-1	R_z	
E_g	2	-1	0	2	-1	0	(R_x, R_y)	$(x^2 - y^2, xy)$ (xz, yz)
A_{1u}	1	1	1	-1	-1	-1		
A_{2u}	1	1	-1	-1	-1	1	z	
E_u	2	-1	0	-2	1	0	(x, y)	

D_{4d}	E	$2S_8$	$2C_4$	$2S_8^3$	C_2	C_2'	$4\sigma_d$		
A_1	1	1	1	1	1	1	1		$x^2 + y^2, z^2$
A_2	1	1	1	1	1	-1	-1	R_z	
B_1	1	-1	1	-1	1	1	-1		
B_2	1	-1	1	-1	1	-1	1	z	
E_1	2	$\sqrt{2}$	0	$-\sqrt{2}$	-2	0	0	(x, y)	
E_2	2	0	-2	0	2	0	0		$(x^2 - y^2, xy)$
E_3	2	$-\sqrt{2}$	0	$\sqrt{2}$	-2	0	0	(R_x, R_y)	(xz, yz)

D_{5d}	E	$2C_5$	$2C_5^2$	$5C_2$	i	$2S_{10}^3$	$2S_{10}$	$5\sigma_d$		
A_{1g}	1	1	1	1	1	1	1	1		$x^2 + y^2, z^2$
A_{2g}	1	1	1	-1	1	1	1	-1	R_z	
E_{1g}	2	j	k	0	2	j	k	0	(R_x, R_y)	(xz, yz)
E_{2g}	2	k	j	0	2	k	j	0		$(x^2 - y^2, xy)$
A_{1u}	1	1	1	1	-1	-1	-1	-1		
A_{2u}	1	1	1	-1	-1	-1	-1	1	z	
E_{1u}	2	j	k	0	-2	$-j$	$-k$	0	(x, y)	
E_{2u}	2	k	j	0	-2	$-k$	$-j$	0		

$j = 2\cos 72°, k = 2\cos 144°$

D_{6d}	E	$2S_{12}$	$2C_6$	$2S_4$	$2C_3$	$2S_{12}{}^5$	$2C_2$	$6C_2{}'$	$6\sigma_d$		
A_1	1	1	1	1	1	1	1	1	1		$x^2 + y^2,\ z^2$
A_2	1	1	1	1	1	1	1	-1	-1	R_z	
B_1	1	-1	1	-1	1	-1	1	1	-1		
B_2	1	-1	1	-1	1	-1	1	-1	1	z	
E_1	2	$\sqrt3$	1	0	-1	$-\sqrt3$	-2	0	0	(x,y)	
E_2	2	1	-1	-2	-1	1	2	0	0		$(x^2 - y^2,\ xy)$
E_3	2	0	-2	0	2	0	-2	0	0		
E_4	2	-1	-1	2	-1	-1	2	0	0		
E_5	2	$\sqrt3$	1	0	-1	$\sqrt3$	-2	0	0	(R_x,R_y)	(xz,yz)

T	E	$4C_3$	$4C_3{}^2$	$3C_2$	$\varepsilon = exp(2\pi i/3)$	
A	1	1	1	1		$x^2+y^2+z^2$
E	1	ε	ε^*	1		$(2z^2-x^2-y^2, x^2-y^2)$
	1	ε^*	ε	1		
T	3	0	0	-1	$(R_x, R_y, R_z);(x,\ y,z)$	$(xz,\ yz,\ xy)$

T_d	E	$8C_3$	$3C_2$	$6S_4$	$6\sigma_d$		
A_1	1	1	1	1	1		$x^2 + y^2 + z^2$
A_2	1	1	1	-1	-1		
E	2	-1	2	0	0		$(2z^2-x^2-y^2, x^2-y^2)$
T_1	3	0	-1	1	-1	(R_x, R_y, R_z)	
T_2	3	0	-1	-1	1	$(x,\ y,z)$	$(xz,\ yz,\ xy)$

O	E	$6C_4$	$3C_2(=C_4{}^2)$	$8C_3$	$6C_2$		
A_1	1	1	1	1	1		$x^2 + y^2 + z^2$
A_2	1	-1	1	1	-1		
E	2	0	2	-1	0		$(2z^2-x^2-y^2, x^2-y^2)$
T_1	3	1	-1	0	-1	$(R_x, R_y, R_z);(x,\ y,z)$	
T_2	3	-1	-1	0	1	$(xz,\ yz,\ xy)$	

O_h	E	$8C_3$	$6C_2$	$6C_4$	$3C_2(=C_4{}^2)$	i	$6S_4$	$8S_6$	$3\sigma_h$	$6\sigma_d$	
A_{1g}	1	1	1	1	1	1	1	1	1	1	$x^2 + y^2 + z^2$
A_{2g}	1	1	-1	-1	1	1	-1	1	1	-1	
E_g	2	-1	0	0	2	2	0	-1	2	0	$(2z^2-x^2-y^2, x^2-y^2)$
T_{1g}	3	0	-1	1	-1	3	1	0	-1	-1	(R_x, R_y, R_z)
T_{2g}	3	0	1	-1	-1	3	-1	0	-1	1	$(xz,\ yz,\ xy)$
A_{1u}	1	1	1	1	1	-1	-1	-1	-1	-1	
A_{2u}	1	1	-1	-1	1	-1	1	-1	-1	1	
E_u	2	-1	0	0	2	-2	0	1	-2	0	
T_{1u}	3	0	-1	1	-1	-3	-1	0	1	1	$(x,\ y, z)$
T_{2u}	3	0	1	-1	-1	-3	1	0	1	-1	

$C_{\infty v}$	E	$2C_\infty{}^\phi$	$2C_\infty{}^{2\phi}...$		$\infty\sigma_v$		
$\Sigma^+ (= A_1)$	1	1	1	...	1	z	$x^2 + y^2, z^2$
$\Sigma^- (= A_2)$	1	1	1	...	-1	R_z	
$\Pi (= E_1)$	2	$2cos\phi$	$2cos2\phi$	...	0	$(x,\ y); (R_x, R_y)$	(xz, yz)
$\Delta (= E_2)$	2	$2cos2\phi$	$2cos4\phi$	...	0		$(x^2 - y^2, xy)$
$\Phi (= E_3)$	2	$2cos3\phi$	$2cos6\phi$	...	0		

$D_{\infty h}$	E	$2C_\infty{}^\phi$	...	$\infty\sigma_v$	i	$2S_\infty{}^\phi$	...	$\infty\sigma_v$		
$\Sigma_g{}^+$	1	1	...	1	1	1	...	1		$x^2 + y^2, z^2$
$\Sigma_g{}^-$	1	1	...	-1	1	1	...	-1	R_z	
Π_g	2	$2cos\phi$	...	0	2	$-2cos\phi$	...	0	(R_x, R_y)	(xz, yz)
Δ_g	2	$2cos2\phi$	...	0	2	$2cos\phi$	...	0		$(x^2 - y^2, xy)$
...		...	...	...		...	...	...	...	...
$\Sigma_u{}^+$	1	1	...	1	-1	-1	...	-1	z	
$\Sigma_u{}^-$	1	1	...	-1	-1	-1	...	1		
Π_u	2	$2cos\phi$	...	0	-2	$2cos\phi$	...	0	$(x,\ y)$	
Δ_u	2	$2cos2\phi$	...	0	-2	$-2cos\phi$	...	0		

K	E	C_4	C_2	C_3	l[a]
s	1	1	1	1	0
p	3	1	-1	0	1
d	5	-1	1	-1	2
f	7	-1	-1	1	3
g	9	1	1	0	4

The character for rotation by an angle ω is given by $\chi(\omega)$
$= sin(l+1/2)\omega/sin(\omega/2)$.
a. These are values for the l quantum number appropriate for $s, p,...$ etc.

1.2. Selected Valence Shell Ionization Potentials (*VSIP*) And Slater
      Exponents For Some Main Group Atoms[b].

Atom	Principal Quantum Number	Slater Exponents		VSIP(eV)		
		s and p	d	s	p	d
H	1	1.3		-13.6		
Li	2	0.650		-5.4	-3.5	
Be	2	0.975		-10.0	-6.0	
B	2	1.300		-15.2	-8.5	
C	2	1.625		-21.4	-11.4	
N	2	1.95		-26.0	-13.4	
O	2	2.275		-32.3	-14.8	
F	2	2.600[a]		-40.0	-18.1	
Na	3	0.733		-5.1	-3.0	
Mg	3	0.950		-9.0	-4.5	
Al	3	1.167		-12.3	-6.5	
Si	3	1.383	1.383	-17.3	-9.2	-6.0
P	3	1.600	1.400	-18.6	-14.0	-7.0
S	3	1.817	1.500	-20.0	-13.3	-8.0
Cl	3	2.033	2.033	-30.0	-15.0	-9.0

a. A value of 2.425 is more often used.
b.*Introduction to Applied Quantum Chemistry*, S. P. McGlynn, L. G.
Vanquickenborne, M. Kinoshita and D. G. Carroll, Holt, Rinehardt and
Winston (1972).

1.3. Selected Valence Shell Ionization Potentials (*VSIP*) For Some Transition Metal Atoms[2].

Atom	VSIP(eV)		
	s	p	d^a
Sc	-8.87	-2.75	-8.51
Ti	-8.97	-5.44	-10.81
V	-8.81	-5.52	-11.0
Cr	-8.66	-5.24	-11.22
Mn	-9.75	-5.89	-11.67
Fe	-9.10	-5.32	-12.6
Co	-7.8	-3.8	-9.7
Ni	-9.17	-5.15	-13.49
Cu	-11.4	-6.06	-14.0
Zn	-12.41	-6.53	

a. In each case the values correspond to the *ns*, *np* and *(n+1)d* orbitals.

1.4. Angular Dependence[2] Of Some Selected Overlap Integrals Between Central Atom *s*, *p* And *d* Orbitals And Ligand σ And π Orbitals[a].

$$S(s,\sigma) = S_\sigma \qquad\qquad S(z^2,\pi_{//})^b = \sqrt{3}HIS_\pi$$
$$S(s,\pi) = 0 \qquad\qquad S(z^2,\pi_\perp) = 0$$
$$S(z,\sigma) = HS_\sigma \qquad\qquad S(x^2 - y^2,\pi_{//}) = -HIS_\pi$$
$$S(z,\pi_{//}) = IS_\pi \qquad\qquad S(x^2 - y^2,\pi_\perp) = 0$$
$$S(z,\pi_\perp) = 0 \qquad\qquad S(xy,\pi_{//}) = 0$$
$$S(z^2,\sigma) = 1/2(3H^2 - 1)S_\sigma \qquad S(xy,\pi_\perp) = IS_\pi$$
$$S(x^2 - y^2,\sigma) = \sqrt{3}/2(F^2 - G^2)S_\sigma \quad S(xz,\pi_{//}) = (I^2 - H^2)S_\pi$$
$$S(xy,\sigma) = \sqrt{3}FGS_\sigma \qquad\qquad S(xz,\pi_\perp) = 0$$
$$S(xz,\sigma) = \sqrt{3}FHS_\sigma \qquad\qquad S(yz,\pi_{//}) = 0$$
$$S(xy,\sigma) = \sqrt{3}GHS_\sigma \qquad\qquad S(yz,\pi_\perp) = HS_\pi$$

a. Here, $\pi_{//}$ and $\pi_\perp$ are ligand π orbitals whose axes lie in a plane containing the z-axis and the ligand and lie parallel and perpendicular to this plane respectively (**1.1**). S_σ and S_π are the overlap integrals between a central atom orbital and a ligand orbital in some reference geometry. They will depend

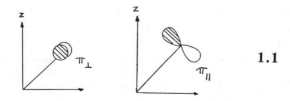

1.1

upon the nature of the atoms concerned, whether the orbital is of s, p or d type and also on the metal-ligand distance. $F = sin\theta cos\phi$, $G = sin\theta sin\phi$, $H = cos\theta$, $I = sin\theta$.
b. The ligand lies in the xz plane.

1.5. Wade's Rules For Main Group Containing Fragments[1,2].

The Number of Skeletal Electrons Contributed By Main Group Containing Fragments[b].

Main Group Element (A)		Cluster Unit[a]	
	A	AH or AY	AH₂ or AY'
Li, Na	[-1]	0	1
Be, Mg, Zn, Cd, Hg	0	1	2
B, Al, Ga, In, Tl	1	2	3
C, Si, Ge, Sn, Pb	2	3	4
N, P, As, Sb, Bi	3	4	5
O, S, Se, Te	4	5	[6]
F, Cl, Br, I	5	[6]	[7]

a. Y is a one electron ligand, such as halogen, Y' is a two electron ligand such as oxygen, *THF* etc.
b. K. Wade, *Adv. Inorg. Chem. Radiochem.*, **18**, 1 (1976).

1.6. Wade's Rules For Transition Metal Containing Fragments[1,2].

The Number of Skeletal Electrons Contributed By Transition Metal Containing Fragments[a].

Transition Metal (M)		Cluster Unit		
	M(CO)₂	M(π-Cp)	M(CO)₃	M(CO)₄
Cr, Mo, W	[-2]	-1	0	2
Mn, Tc, Re	-1	0	1	3
Fe, Ru, Os	0	1	2	4
Co, Rh, Ir	1	2	3	5
Ni, Pd, Pt	2	3	4	6

a.   K. Wade, *Adv. Inorg. Chem. Radiochem.*, **18**, 1 (1976).

1.7.   The First Eight Deltahedra (Trigonal Bipyramid, Octahedron,
        Pentagonal Bipyramid, Dodecahedron, Tricapped Trigonal Prism
        Bicapped Square Antiprism, Undecahedron and Icosahedron.)

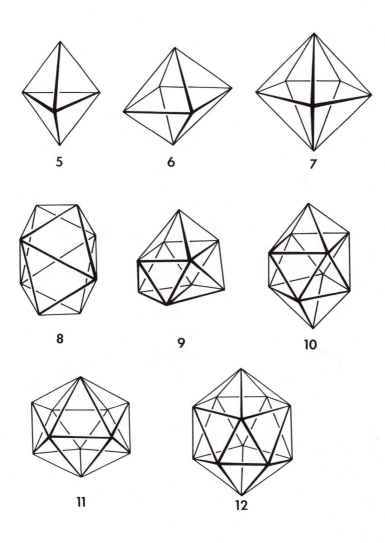

1.2

1.8. Isolobal Relationships[1].

1.3

See too: R. Hoffmann, *Angew. Chemie, Int. Ed. Engl.*, **21**, 711 (1982).

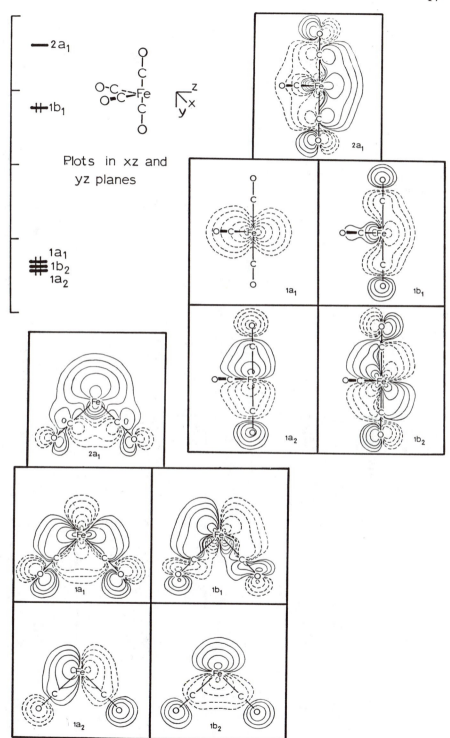

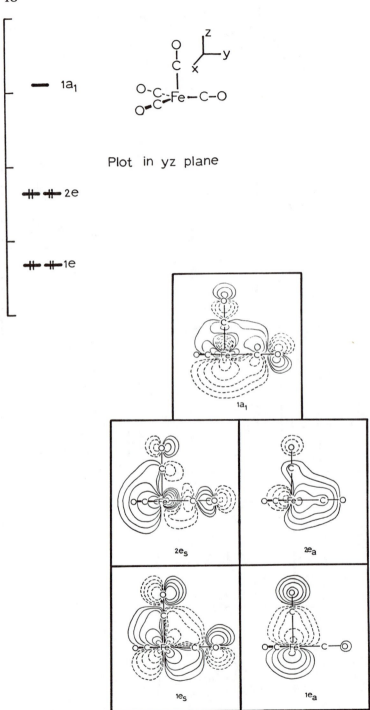

Plot in yz plane

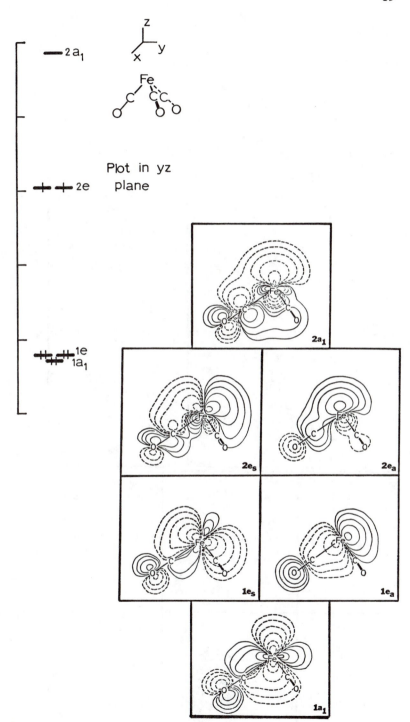

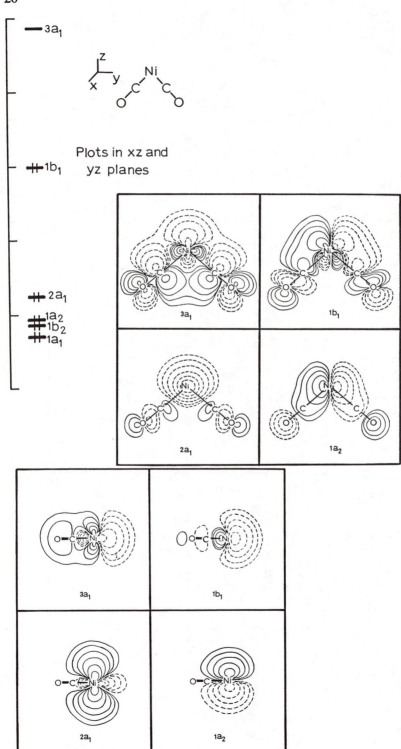

Plots in xz and
yz planes

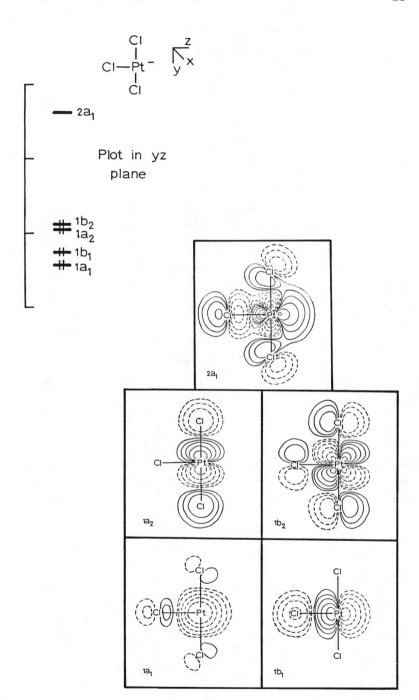

1.4

1.10. The Periodic Table and Electronegativities of The Elements[a].

1	2	3	4	5	6	7	8	9	10	11	12	13	14	15	16	17	18
H 2.300																	
Li 0.912	Be 1.576											B 2.051	C 2.544	N 3.066	O 3.610	F 4.193	Ne 4.787
Na 0.869	Mg 1.293											Al 1.613	Si 1.916	P 2.253	S 2.589	Cl 2.869	Ar 3.242
K 0.734	Ca 1.034	Sc 1.2	Ti 1.3	V 1.5	Cr 1.6	Mn 1.6	Fe 1.6	Co 1.7	Ni 1.8	Cu 1.8	Zn 1.7	Ga 1.756	Ge 1.994	As 2.211	Se 2.424	Br 2.685	Kr 2.966
Rb 0.706	Sr 0.963	Y 1.1	Zr 1.2	Nb 1.2	Mo 1.3	Tc 1.4	Ru 1.4	Rh 1.4	Pd 1.4	Ag 1.4	Cd 1.5	In 1.656	Sn 1.824	Sb 1.984	Te 2.158	I 2.359	Xe 2.582
Cs 0.69	Ba 0.89	La 1.1	Hf 1.2	Ta 1.3	W 1.4	Re 1.5	Os 1.5	Ir 1.6	Pt 1.4	Au 1.4	Hg 1.4	Tl 1.36	Pb 1.61	Bi 2.06			

Ln 1.1

a. Main Group Values from L. Allen, *J. Amer. Chem. Soc.*, **111**, 9003 (1989).

1.11. The Hückel Energies of Some Small Systems[a].

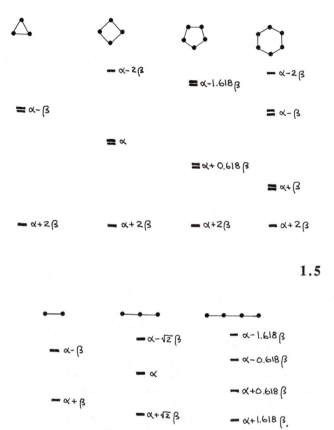

1.5

a. *The HMO Model and its Application*, E. Heilbronner and H. Bock, Verlag Chemie (1976).

1.12. Perturbation Theory.

Many of the exercises in this book utilize perturbation theory to build up more complicated orbital patterns from that of a readily accessible system, or to develop them within the context of reduced symmetry. We use a form of perturbation theory developed in reference 1 which the reader should consult for details. We briefly review the results here.

The orbitals of the starting, unperturbed system are given by a normal LCAO expansion:

$$\Psi^{\circ}_i = \sum_{\mu} c^{\circ}_{\mu i} \chi_{\mu} \tag{1}$$

where χ_{μ} are atomic orbitals and the $c^{\circ}_{\mu i}$ are the mixing coefficients in the molecular orbital Ψ°_i. The orbitals are assumed to be orthonormal. The associated eigenvalues for the Ψ°_i are given by:

$$\mathcal{H}^{eff}\Psi^\circ_i = e^\circ_i \Psi^\circ_i \tag{2}$$

where $\mathcal{H}^{eff}$ is some effective Hamiltonian. For convenience we will use a one-electron model, and so $\mathcal{H}^{eff}$ contains no electron-electron terms. Here, and in the following discussion, Greek subscripts index atomic orbitals and Roman subscripts molecular orbitals. The orbitals of the perturbed system are written in terms of the orbitals of the unperturbed system as:

$$\Psi_i = t_{ii}\Psi^\circ_i + \sum_{j\neq i} t_{ji} \Psi^\circ_j \tag{3}$$

where

$$t_{ii} = t_{ii}(1) + t_{ii}(2) + \ldots$$

$$t_{ji} = t_{ji}(1) + t_{ji}(2) + \ldots \tag{4}$$

and

$$e_i = e_i^\circ + e_i(1) + e_i(2) + \ldots \ . \tag{5}$$

Here the superscripts (1), (2) refer to the first and second order corrections respectively to the unperturbed system. As a general rule of thumb, the magnitude of the first order corrections to the wavefunction are much smaller than the coefficients for the zeroth order, unperturbed solution. Similarly the second order orbital and energy corrections are smaller than the first order ones. The mixing coefficient t_{ji} refers to the mixing of Ψ°_j into Ψ°_i.

When the perturbation is turned on the molecular orbitals need not be orthogonal any longer. In general we may write the overlap between the molecular orbitals, $\tilde{S}_{ij}$, as;

$$<\Psi^\circ_i/\Psi^\circ_j> = \tilde{S}_{ij} = \sum_\mu \sum_\nu c^\circ_{\mu i} \delta S_{\mu\nu} c^\circ_{\nu j} \tag{6}$$

The value of $\delta S_{\mu\nu}$ measures the change in the overlap integral between two atomic orbitals χ_μ and χ_ν in the two molecular orbitals Ψ°_i and Ψ°_j. Likewise there will be an interaction integral between the two molecular orbitals of the form;

$$<\Psi^\circ_i/\mathcal{H}^{eff}/\Psi^\circ_j> = \tilde{\Delta}_{ij} = \sum_\mu \sum_\nu c^\circ_{\mu i} \delta H_{\mu\nu} c^\circ_{\nu j} \tag{7}$$

Using these orbital parameters the energy changes may be simply written as;

$$e_i(1) = \tilde{\Delta}_{ii} \tag{8}$$

$$e_i^{(2)} = \sum_j \tilde{\Delta}_{ij}^2 / (e_i^\circ - e_j^\circ) \qquad (9)$$

There is a direct relationship between the magnitude of the overlap and interaction integrals introduced by the perturbation, namely $\tilde{\Delta}_{ij} \propto -\tilde{S}_{ij}$. Thus;

$$e_i^{(2)} \propto \sum_j \tilde{S}_{ij}^2 / (e_i^\circ - e_j^\circ) \qquad (10)$$

The mixing coefficients turn out to be;

$$t_{ji}^{(1)} \propto \sum_j -\tilde{S}_{ij} / (e_i^\circ - e_j^\circ) \qquad (11)$$

and

$$t_{ki}^{(2)} \propto \sum_j \tilde{S}_{ij}\tilde{S}_{jk} / (e_i^\circ - e_k^\circ)(e_i^\circ - e_j^\circ) \qquad (12)$$

There are three different types of perturbation that are commonly encountered in electronic structure problems. They are of the intermolecular, geometry and electronegativity types and are now discussed in turn.

(i)   Intermolecular Perturbation

    This perturbation is used to construct the molecular orbitals and evaluate their energies when the orbitals of two fragments (the unperturbed system) are allowed to interact. An example might be the assembly of a molecular orbital diagram for the ethyl cation from the orbitals of CH_3 and CH_2 fragments.The geometry of each fragment is such that their individual atomic arrangements are unaltered once the new molecule is formed. Let Ψ°_i be associated with one fragment and Ψ°_j with the other. The perturbation is just the switching on of the overlap between the two sets of orbitals on each fragment. Thus it is clear that $e_i^{(1)} = 0$ from equations 7, 8 so that the energy becomes

$$e_i = e_i^\circ + e_i^{(2)} + \dots . \qquad (13)$$

Here $e_i^{(2)}$ is just the sum of the pairwise interactions from equation 10. The form of the final wavefunctions is just;

$$\Psi_i = (1 + t_{ii}^{(2)})\Psi^\circ_i + t_{ji}^{(1)}\Psi^\circ_j + t_{ki}^{(2)}\Psi^\circ_k \qquad (14)$$

where $t_{ji}^{(1)}$ and $t_{ki}^{(2)}$ are given by equations 11, 12, and $(1 + t_{ii}^{(2)}) \approx 1$. In equation 14, Ψ°_i and Ψ°_k are contained in the same fragment. Although the overlap between them is necessarily zero, Ψ°_k can mix into Ψ°_i in second order provided both fragment orbitals have a non-zero overlap with Ψ°_j on the other fragment. This is how, for example, *sp* hybrid orbitals come about in molecular orbital theory. Although *s* and *p* orbitals are orthogonal, they may mix together *via* overlap with a third orbital on another atom. For the graphical display of equation 14 in the answers we have adopted the standard that the first order terms are enclosed by parentheses, while the second order mixing is indicated within brackets.

The above discussion requires that no degeneracies exist between the two fragments, *i.e.*, $e_i° \neq e_j°$ in equations 10, 11. When the two fragments are identical, degenerate perturbation theory must be used for $e_i° = e_j°$. The result is that

$$e_i = e_i° + e_i^{(1)} + e_i^{(2)} + \ldots . \qquad (15)$$

$$e_j = e_j° + e_j^{(1)} + e_j^{(2)} + \ldots . \qquad (16)$$

where $e_i^{(1)} = -e_j^{(1)} \propto \tilde{S}_{ij}$ and $e_i^{(2)} = e_j^{(2)} \propto \tilde{S}_{ij}^2$. The two final molecular orbitals are simply linear combinations of the two fragment orbitals

$$\Psi_{i,j} = (2\pm2\tilde{S}_{ij})^{-1/2}(\Psi_i° \pm \Psi_j°) \qquad (17)$$

Often both degenerate and nondegenerate interactions are present.

(ii) Geometrical Perturbations.

An example of a geometrical perturbation is the pyramidalization of planar ammonia. Invariably we are interested in the way the orbitals of a molecule in a highly symmetric point group change on lowering the symmetry. Now all of the orbitals concerned are associated with one fragment so that;

$$\Psi_i = (1 + t_{ii}^{(2)})\Psi_i° + t_{ji}^{(1)}\Psi_j° \qquad (18)$$

Here $(1 + t_{ii}^{(2)}) \approx 1$ as before and $t_{ji}^{(1)}$ is given by equation 11. The orbital energies become;

$$e_i = e_i° + e_i^{(1)} + e_i^{(2)} + \ldots . \qquad (19)$$

where the first order correction $(e_i^{(1)} \propto -\tilde{S}_{ij})$ is often important since the distortion invariably leads to a change in the overlap integrals between the atomic orbitals within the molecule. The second order correction is given by equation 10.

(iii) Electronegativity Perturbation

The substitution of an atom(s) in a molecule will also create changes in the wavefunctions and orbital energies. The Coulomb integral associated with χ_α in the unperturbed molecule is $<\chi_\alpha|\mathcal{H}^{eff}|\chi_\alpha> = H_{\alpha\alpha}$. With a change in electronegativity the new Coulomb integral is just $H_{\alpha\alpha} + \delta\alpha$, where $\delta\alpha < 0$ if the electronegativity of the atom is increased and $\delta\alpha > 0$ if the electronegativity is decreased. Then;

$$e_i(1) = (c°_{\alpha i})^2 \delta\alpha \qquad (20)$$

$$e_i(2) = \sum_j (c°_{\alpha i} c°_{\alpha j} \delta\alpha)^2 / (e_i° - e_j°) \qquad (21)$$

and the resulting wavefunction are evaluated as in equation 18 with

$$t_{ji}(1) \propto \sum_j c°_{\alpha i} c°_{\alpha j} \delta\alpha / (e_i° - e_j°) \qquad (22)$$

This result uses the assumptions that the geometry does not change as a result of the atomic substitution, and that the overlap integrals between the atomic orbitals are unaffected by the substitution.

Chapter II.

General

2.1. Construct correctly normalized wavefunctions for the bonding and antibonding orbitals of the H_2 molecule. Use the simplest model, including overlap, set $<i/\mathcal{H}^{eff}/j> = \alpha$ $(i = j)$ and $= \beta$ $(i \neq j)$ to determine the orbital energies. Add electrons appropriate for the He_2 species and show that the He-He potential is only repulsive if overlap is included in the model. Using a similar model for the π orbitals in ethylene, show how substituted ethylenes of the type $RR'C=CRR'$ undergo cis-trans isomerism upon photochemical excitation.

*2.2. The Jahn-Teller theorem specifically excludes the case of linear molecules. (a) Show that the symmetric direct product of e_k is $a_1 + e_{2k}$ in linear groups ($e_1 = \pi$, $e_2 = \delta$, $e_3 = \phi$ etc.) (b) Satisfy yourself that all bending modes of linear molecules are of $e_1 = \pi$ symmetry. (c) Use these results to understand this exclusion.

2.3. Write down the wavefunctions for the four equivalent sp^3 hybrids derived from the $2s$ and $2p$ orbitals of carbon. Show that these orbitals are not orthogonal and thus cannot be solutions of the Schrödinger wave equation for tetrahedral molecules such as CH_4.

*2.4. **2.1** shows two possible geometries for the hypothetical molecule H_6. In **a**, $r_1 = r_2$, etc. = 1.0Å; i.e., all H-H-H angles are 120.0°. In **b**, $r_1 = 0.74$Å and $r_2 = 4.00$Å; i.e., the distances alternate. For reference, the H-H distance in molecular H_2 is 0.74Å. Carefully draw an orbital correlation diagram connecting **a** to **b**. Explain in detail why the orbitals go up, down, or remain at the same energy. Hint: consider very carefully the geometry in **b**; some of the orbitals have a very different shape when compared to those in **a**; why? Finally, which of the two structures is expected to be of lower energy and why?

2.1

28

2.5.   Draw a qualitative sketch of $S_{\mu\nu}$ as a function of distance and angle for each of the situations shown in **2.2**.

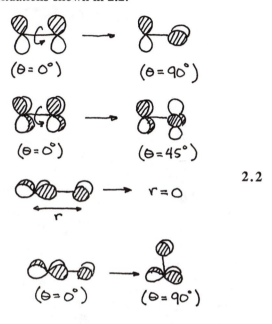

2.2

2.6.   Use perturbation theory to work out the energetics of the situations of **2.3**. Show the orbitals and relative energies.

2.7.   This problem investigates the noncrossing rule for a generalized situation. Consider two orbitals, χ_1 and χ_2, whose orbital energies vary with respect to some distortion, x, according to the following:

$$H_{11} = -5 + x, H_{22} = 5 - x$$

Here x runs from 0 to 10, and in the absence of an interaction, the two orbitals would cross at $x = 5$. However, let us assume that χ_1 and χ_2 have the same symmetry and, therefore, they will interact with each other *via* a matrix element $H_{12} = 1$ as in **2.4**. Calculate the two resulting energy levels, E^+ and E^-, in the presence of this interaction for $x = 0$ to 10 with unit intervals, assuming that the overlap between the two basis functions, $S_{12} = 0$. Plot the results and calculate the corresponding wavefunctions:

$$\Psi^+ = c_1^+\chi_1 + c_2^+\chi_2 \text{ and } \Psi^- = c_1^-\chi_1 + c_2^-\chi_2$$

How do the wavefunctions change along the coordinate x? What would happen to the energies and wavefunctions if H_{12} were 0.01 instead of 1?

2.8.   A study of the *NMR* spectra of paramagnetic molecules provides valuable clues as to the spatial location of the unpaired electrons. The nuclei

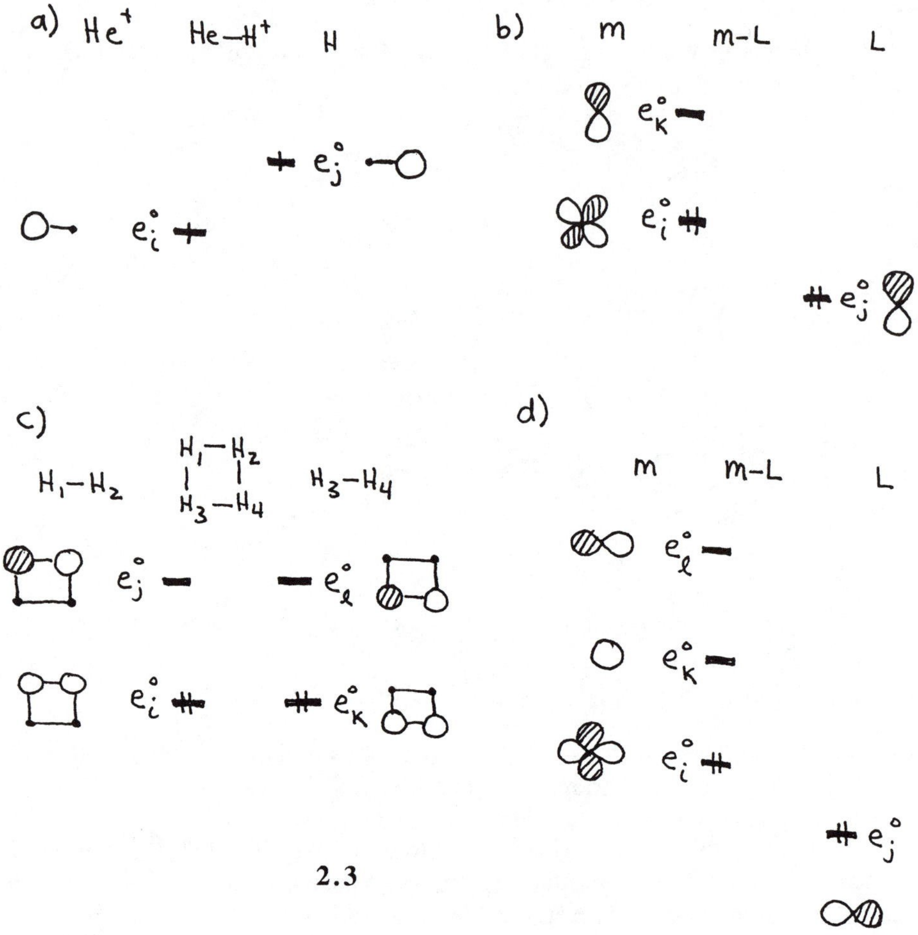

a) He^+ $He-H^+$ H

b) m $m-L$ L

c)

H_1-H_2 $\begin{matrix} H_1-H_2 \\ | \quad | \\ H_3-H_4 \end{matrix}$ H_3-H_4

d) m $m-L$ L

2.3

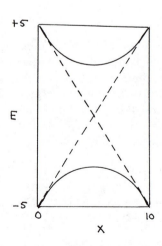

2.4

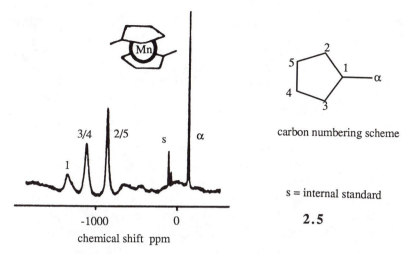

3/4   2/5   S   α

1

-1000   0

chemical shift ppm

carbon numbering scheme

s = internal standard

2.5

which experience the largest paramagnetic shift are those where the unpaired density is highest. **2.5** shows the ^{13}C NMR spectrum of the molecule *(MeCp)₂Mn* in *THF*. Show, by using the molecular orbital diagram you derived in Question 5.23 how the experimental results may be understood.

2.9.   If CH_4 is monosubstituted by deuterium the point symmetry naturally changes. Identify the groups involved. Generate a molecular orbital diagram for the molecule using the correct symmetry labels for the two point groups.

*2.10‡.   One of the problems, often difficult to tackle using qualitative molecular orbital ideas, is the energetic comparison of molecules where the atoms have different coordination numbers. For example one could envisage the two structures of **2.6** for the H_5 molecule and its ions, but how good are simple energetic comparisons between the two using Hückel theory? One way

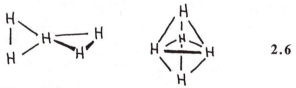

2.6

around this problem (which sometimes works) is to compare the energies of the two structures using the constraint that the second moment (the sum of the squares) of the energy levels is set equal for the two structures. Use a program which will generate the eigenvalues and eigenvectors of a real symmetric matrix to construct Hückel molecular orbital diagrams for the two structures of **2.6**. Scale the energy levels of one of the molecules such that the two sets have the same second moment and predict the geometries of H_5^+ and H_5^-.

*2.11.   Use the character table for the Group *K* (Kügelgruppe) given in Chapter I to determine the symmetry species and spin states of the electronic states arising from the following atomic configurations.   (i) s^2, (ii) p^2, (iii) d^2 (iv) d^8.

2.12. Use perturbation theory (and group theory considerations) to determine the molecular orbitals of an MO molecule where M is a transition metal. For this purpose, use the metal $3d$ and $4s$ atomic orbitals and choose z as the $M\text{-}O$ axis. Also remember that the $3d$ atomic orbitals lie lower in energy than $4s$. For oxygen, use only p orbitals, labeling them x, y and z. The electronegativities of the transition metals and oxygen are given in Table 1.10. Write out the formulae for the energy corrections and wavefunctions in each case. Draw out an orbital interaction diagram and draw out the form of each resultant molecular orbital.

2.13. Assemble molecular orbital diagrams for linear and equilateral triangular H_3 molecules. Evaluate the energies of the orbitals by defining an interaction energy (β) between adjacent $1s$ orbitals. Determine the symmetry species of the ground electronic state of the ions $H_3^{\pm}$ and H_3, assuming of course they have the same geometry as the cation. Hence predict the geometries of H_3^{+} and H_3^{-}.

2.14. (a) Construct a molecular orbital diagram for the methane molecule, using $2s$ and $2p$ orbitals on carbon and $1s$ orbitals on hydrogen. Explain the principles of construction as you go along, *i.e.*, why do the molecular orbitals have the positions you draw for them? (b) Show the connection between the Lewis structure of methane and your diagram. Where are the electron-pair bonds? Is there any sp hybridization in the molecule? (c) Interpret the photoelectron spectrum of methane (**2.7**) in terms of your molecular orbital model. Is it possible to use the traditional organic chemists sp^3 hybrid model to do this too?

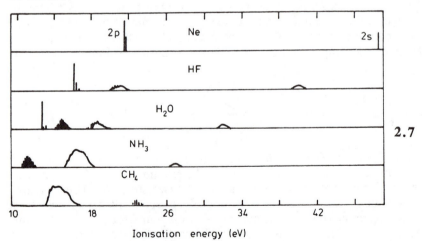

Ionisation energy (eV)

2.7

*2.15. Do a simple molecular orbital calculation for the $Li\text{-}H$ molecule. For this calculation consider only the $1s$ orbital on hydrogen, χ_1, and a $2s$ orbital on lithium, χ_2, *i.e.*, assume that the two $1s$ electrons on lithium are core-like and do not contribute to chemical bonding. The relevant H_{ii} values come from Table 1.2. The value of $\langle \chi_1 | \chi_2 \rangle = S_{12}$, given a $Li\text{-}H$ distance of 3.015 bohr (in atomic units), is 0.3609. For the resonance integral, H_{12}, use the Wolfsberg-Helmholz approximation:

$$H_{12} = (1/2)K\,(H_{11}+H_{22})\,S_{12}$$

where $K = 1.75$.

(a) Set up the secular determinant for this problem and solve for the eigenvalues, E_1 and E_2 for the mixing coefficients in the new wavefunctions ψ_1 and ψ_2.

(b) Do a Mulliken population analysis on *Li-H* using your answer in (a) and compute the charges on the atoms, Q_i, and the *Li-H* bond overlap population P_{Li-H}. The dipole moment, μ^* in Debyes, can be simply computed for a neutral diatomic molecule by using a point charge approximation as:

$$\mu^* = 2.54Qr$$

where Q is the charge on one of the atoms and r is the distance between the two atoms in bohrs. Compute the dipole moment for *LiH* using your charges.

The results of two *ab initio SCF* molecular orbital calculations on *LiH* are now described. Here, the *1s* level on lithium is explicitly included in the calculation and no approximations have been made in terms of the integrals. The only "parameterization" is introduced in the basis set. In the first calculation, a minimal basis set is used. This means that there is essentially one Slater type function for each atomic orbital type, *i.e.*, *1s, 2s, 2p* for lithium and *1s* for hydrogen. The second calculation uses an extended basis set where the *1s* core of lithium still is represented by one function but the *2s* and *2p* atomic orbitals are each represented by three functions and the same is true for the *1s* atomic orbital of hydrogen. Furthermore, there are *d* polarization functions on lithium and *p* polarization functions on hydrogen. Listed in Table 2.1 are the computed charges, Q, on the atoms, the total Mulliken overlap population between *Li* and *H*, P_{Li-H}, the dipole moment computed within the point charge approximation, μ^*, and the computed dipole moment which does not employ the point charge approximation, μ.

(c) Describe why your answers for Q, P_{Li-H} and μ^* are different from those at the minimal basis set level and what happens to these numbers on going from the minimal to extended basis level. Note, this has nothing to do with the core

Table 2.1

Property	Minimal Basis	Extended Basis
Q_H	-0.2194	-0.3569
Q_{Li}	+0.2194	+0.3569
P_{LiH}	0.7556	0.7428
μ^*	1.68	2.73
μ	5.89	6.00
$E_1(eV)$	-8.07	-8.14

$1s$ shell on lithium.

(d) Describe in physical terms why μ^* is much less than μ for the two *ab initio* calculations and why the two *ab initio* values of E_1, listed in Table 2.1, are greatly different from what you calculated for E_1.

2.16. (a) Normally, as an atom becomes more electronegative, bond lengths to it become shorter. A typical series for *HA* molecules is shown in Table 2.2. Why does this occur? Likewise, the bond dissociation energy for homolytic fission (*D.E.*) also increases. Why?

Table 2.2

HA	Length (Å)	D.E. (kcal/mol)
BeH	1.30	53
BH	1.23	70
CH	1.12	80
NH	1.04	85
OH	0.97	102
FH	0.92	135

(b) While the statements in (a) are generally true, there can be exceptions. Describe why the bond lengths and *D.E.* vary the way they do for the series given in Table 2.3.

Table 2.3

molecule	Length (Å)	D.E. (kcal/mol)
C_2	1.24	144
CN	1.17	188
CO	1.13	256
CF	1.27	131

(c) Describe why the bond lengths and *D.E.* vary the way they do for the molecules listed in Table 2.4.

Table 2.4

molecule	Length (Å)	D.E. (kcal/mol)
CO	1.13	256
CS	1.53	174
SiO	1.51	183
SiS	1.93	148
SnO	1.84	127
SnS	2.21	110

2.17. Calculate the bond overlap populations in H_2, H_2^+ and equilateral triangular H_3^+ assuming the same H-H distance in each case. The *H-H* distance in molecular H_2 is 0.74Å and in H_2^+ it is 1.06Å. Use these data and your computed result to make a suggestion concerning the expected *H-H* distance in H_3^+.

2.18. *Ab initio* calculations have indicated that the most stable structures for HeH_3^+ and BeH_3^+, although similar, differ in the location of the heavy atom (**2.8.**). Construct a molecular orbital diagram for the all-hydrogen molecule and use electronegativity perturbation theory to rationalize the difference.

2.8

2.19. Listed in Table 2.5 are the eigenvalues (in eV) and wavefunctions for CO using the coordinate system shown. (a) Draw out cartoons of all the orbitals. This is an *ab initio* calculation which used a single zeta basis set. Compare these wavefunctions with the approximate ones which you derive from an electronegativity perturbation of the homoatomic molecule. Note that the core, $1s$, electrons are included in the *ab initio* calculation. (b) Normally, one would draw the formula for CO as: $\cdot C\equiv O^+$. Listed in Table 2.6 are four *ab initio* calculations for CO using popular basis sets of increasing sophistication. Listed is the total energy for each calculation, E_{TOT}, in hartrees, the charge on carbon, Q_C, the charge on oxygen, Q_O, and the overlap population, P_{CO}. What does this information tell us about the Mulliken population analysis and the "reality" of charges?

Table 2.6.

basis set	E_{TOT}	Q_C	Q_O	P_{CO}
1) *single zeta*	-111.225	+0.201	-0.201	1.061
2) *double zeta*	-112.552	+0.394	-0.394	0.790
3) *triple zeta + d*	-112.768	+0.201	-0.201	1.225
4) *triple zeta + d + f*	-112.773	+0.159	-0.159	1.280

Note: Basis sets 3) and 4) are similar except that f orbitals have been added to C and O in basis set 4).

Table 2.5

	orbital number									
	1	2	3	4	5	6	7	8	9	10
Energy (eV)	-551.4580	-299.522	-39.423	-18.8861	-14.8633	-14.8633	-12.0548	8.4643	8.4643	28.0917
Carbon atom coefficients										
$1s$	0.0004	0.9936	0.1266	-0.1660	0.0000	0.0000	-0.1671	0.0000	0.0000	0.1219
$2s$	-0.009	0.0283	-0.2403	0.5384	0.0000	0.0000	0.7611	0.0000	0.0000	-0.9741
$2p_x$	0.0000	0.0000	0.0000	0.0000	0.0000	-0.4455	0.0000	0.0000	0.9316	0.0000
$2p_y$	0.0000	0.0000	0.0000	0.0000	-0.4455	0.0000	0.0000	0.9316	0.0000	0.0000
$2p_z$	0.0074	-0.0070	0.1670	-0.0679	0.0000	0.0000	0.5728	0.0000	0.0000	1.2292
Oxygen atom coefficients										
$1s$	0.9942	-0.0002	-0.2224	0.1320	0.0000	0.0000	0.0011	0.0000	0.0000	-0.1283
$2s$	0.0278	-0.0069	-0.7663	-0.6459	0.0000	0.0000	0.0380	0.0000	0.0000	1.0857
$2p_x$	0.0000	0.0000	0.0000	0.0000	0.0000	-0.7909	0.0000	0.0000	-0.6641	0.0000
$2p_y$	0.0000	0.0000	0.0000	0.0000	-0.7909	0.0000	0.0000	-0.6641	0.0000	0.0000
$2p_z$	0.0068	-0.0001	0.2181	0.6279	0.0000	0.0000	-0.4339	0.0000	0.0000	0.9693

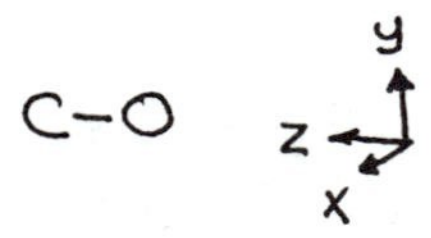

C—O

y, x, z (axis diagram)

Answers

2.1. Let us write an arbitrary wavefunction $\psi = a\phi_1 + b\phi_2$ for the H_2 molecule where ϕ_1 and ϕ_2 are the $1s$ orbitals on atoms 1 and 2 respectively. The square of this function $\psi^2 = a^2\phi_1^2 + b^2\phi_2^2 + 2ab\phi_1\phi_2$ describes the electron distribution. Since the hydrogen atoms *1* and *2* are symmetry equivalent then the electron density located at each center must be the same. This sets a strong restriction on the values of a and b since $a^2 = b^2$ or $a = \pm b$. Thus symmetry tells us that there are only two solutions for this orbital problem which may be written as:

$$\psi_b = N_b(\phi_1 + \phi_2)$$
$$\psi_a = N_a(\phi_1 - \phi_2)$$

where $\psi_{b,a}$ are bonding and antibonding orbitals respectively, and $N_{b,a}$ are normalization constants. To properly normalize these orbitals we need to use the condition that $\int\psi^2 d\tau = 1$. So:

$$\int\psi_{b,a}^2 d\tau = N_{b,a}^2(\int\phi_1^2 d\tau + \int\phi_2^2 d\tau \pm 2\int\phi_1\phi_2 d\tau).$$

or using Dirac notation

$$|<a \text{ or } b|a \text{ or } b>|^2 = N_{b,a}^2(|<1|1>|^2 + |<2|2>|^2 \pm 2<1|2>).$$

If $\int\phi_1\phi_2 d\tau = <1|2>$ is the overlap integral, S_{12}, between the orbitals ϕ_1 and ϕ_2, then $N_{b,a}^2 = 1/\sqrt{2}(1 \pm S_{12})$. Now to calculate the energy of each orbital we need to evaluate $e_i = \int\phi_i \mathcal{H}^{eff}\phi_i d\tau = <i|\mathcal{H}^{eff}|i>$, where $\mathcal{H}^{eff}$ is some effective Hamiltonian. For ψ_b, in Dirac notation:

$$e_b = N_b^2(<1|\mathcal{H}^{eff}|1> + <2|\mathcal{H}^{eff}|2> + 2<1|\mathcal{H}^{eff}|2>).$$

If we write $<1|\mathcal{H}^{eff}|1>$ as the energy of an electron in orbital 1 ($= \alpha$) and $<1|\mathcal{H}^{eff}|2>$ as the interaction energy between the two orbitals 1 and 2 ($= \beta$), then

$$e_b = (2\alpha + 2\beta)/(2(1 + S_{12}))$$
$$= (\alpha + \beta)/(1 + S_{12})$$

Similarly

$$e_a = (\alpha - \beta)/(1 - S_{12}).$$

Importantly, the antibonding orbital is destabilized more than the bonding orbital is stabilized as shown in **2.9**.

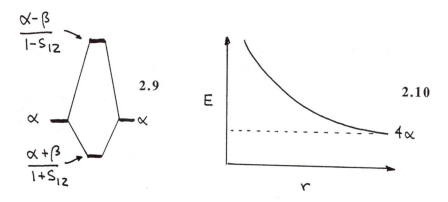

2.9

2.10

The energy of the He_2 molecule with the electronic configuration $\psi_b^2 \psi_a^2$ is simply evaluated as $4(\alpha - \beta S_{12})/(1 - S_{12}^2)$. At large internuclear separations β, $S_{12} \sim 0$ and the total energy $\sim 4\alpha$. Since $\beta < 0$, the molecule becomes less stable relative to two isolated He atoms as the internuclear separation (R) decreases with a consequent decrease in S_{12} (**2.10**). If $S_{12} = 0$ however, there is no change in energy.

The molecular orbital diagram for the π-system of ethylene is very similar to that of **2.9**, except that α, β are now appropriate for carbon $p\pi$ orbitals. In the electronic ground state the total energy is just $2(\alpha + \beta)/(1 + S_{12})$ as before. But let us see how this changes on rotation around the C-C

2.11

bond. Notice in **2.11** that at $\theta = 90°$ (see problem 2.5) there is no overlap at all between the two $p\pi$ orbitals so that at this point β and $S_{12} = 0$. **2.12** shows how the energies of ψ_a and ψ_b vary with θ, and **2.13** the variation in the total energy of the ground and first excited singlet states. The pathway followed on photochemical excitation is indicated in **2.13** by a dashed line. It involves a twist to the $\theta = 90°$ structure followed by a return to the $\theta = 0°$ structure. Geometrically this is shown in **2.14**. Clearly such a motion can lead to *cis-trans* isomerization which in the absence of competing effects would have a quantum yield of 0.5.

The He_2 and ethylene molecules just simply differ in the way the instability of equal occupation of ψ_a and ψ_b may be relieved. In the former, the only coordinate possible is the one which separates the two atoms, but in the latter there is an internal motion which may be employed. There are many two orbital-four electron problems of this type where an energetically unfavorable interaction is relieved on distortion.

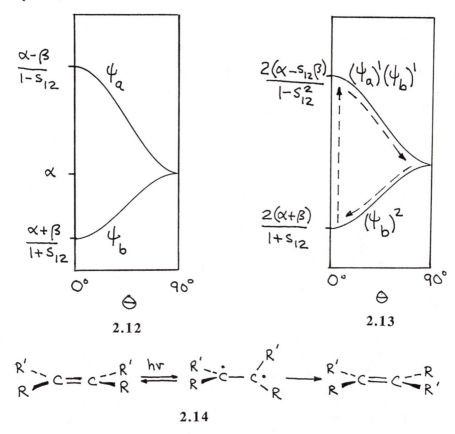

2.12   2.13

2.14

2.2. The first order Jahn-Teller instability for the state $|i>$ along a displacement coordinate q, is associated with a non-zero value of the integral $<i/(\partial\mathcal{H}/\partial q)_0/i>$. Group theory gives access to the allowed symmetry species of q. For this integral to be non-zero Γ_q must be contained in the symmetric direct product of Γ_i. The latter is easy to evaluate for the $C_{\infty v}$ point group. A part of the character table is shown in Table 1.1.

(a) The character of the symmetric direct product is just $\chi^2_{sym}(\mathcal{R}) = (\chi^2(\mathcal{R}) + \chi(\mathcal{R}^2))/2$. This is evaluated in Table 2.7.

(b) All the bending vibrations involve atomic displacements along x and y (if $C_\infty\phi$ lies along z). We can see how x and y behave under a $C_\infty\phi$ rotation in **2.15**. The transformation properties are just

$$x' = cos\phi \cdot x + sin\phi \cdot y$$
$$y' = -sin\phi \cdot x + cos\phi \cdot y$$

The character associated with this transformation matrix is just the sum of the diagonal elements, namely $2cos\phi$. Thus all such bending motions transform as $\pi (= e_1)$.

Table 2.7.

for e_k	E	$2C_\infty\phi$	$2C_\infty 2\phi$	...	$\infty\sigma_v$
$\chi(\mathcal{R})$	2	$2\cos k\phi$	$2\cos 2k\phi$	...	0
$\chi(\mathcal{R}^2)$	2	$2\cos 2k\phi$	$2\cos 4k\phi$	...	2
$\chi^2(\mathcal{R})$	4	$4\cos^2 k\phi$	$4\cos^2 2k\phi$	...	0
$\chi^2_{sym}(\mathcal{R})$	3	$\cos 2k\phi + 2\cos^2 k\phi$	$\cos 4k\phi + 2\cos^2 2k\phi$	...	1
"	3	$2\cos 2k\phi + 1$	$2\cos 4k\phi + 1$	...	1

Thus $\Gamma_i^2(sym)$ simply reduces to $a_1 + e_{2k}$.

2.15

(c) Since the integral $<i/(\partial\mathcal{H}/\partial q)_0/i>$ is only non-zero if Γ_q is contained in $a_1 + e_{2k}$, it will always be zero since $\Gamma_q = e_1$. This does not mean that degenerate electronic states of linear molecules are geometrically stable (except for the special case of diatomic molecules). Singlet CH_2, for example, with an electronic configuration at the linear geometry of $(\pi_u)^2$, has a bent geometry which we may regard as arising *via* a second order Jahn-Teller distortion. The water molecule, which does not have a degenerate electronic state at the linear geometry, $(\pi_u)^4$, also has a bent geometry for the same reason. For further reading see refs 1,2.

2.3   The four classic Pauling sp^3 hybrids are simply given by

$$\phi_1 = 1/2(s - p_x + p_y - p_z)$$
$$\phi_2 = 1/2(s - p_x - p_y + p_z)$$
$$\phi_3 = 1/2(s + p_x - p_y - p_z)$$
$$\phi_4 = 1/2(s + p_x + p_y + p_z)$$

The energy of each hybrid is just

$$1/4<(s + p_x + p_y + p_z)/\mathcal{H}^{eff}/(s + p_x + p_y + p_z)>$$
$$= 1/4(e_s + 3e_p)$$

and the interaction energy between any pair is

$$1/4<(s + p_x + p_y + p_z)/\mathcal{H}^{eff}/(s - p_x - p_y + p_z)>$$
$$= 1/4(e_s - e_p)$$

Since one requirement for a set of stationary states of the Schrödinger wave equation is that they are orthogonal, orbitals built out of these hybrids cannot be eigenstates of the molecule. We know too of course, that in tetrahedral molecules with maximal symmetry T_d, no degeneracy higher than three may exist. (The molecular orbital diagram for CH_4 is derived in Question 2.14.)

2.4. There are several very important features in the solution to this problem. The structure of **b** is nothing more than three very weakly interacting H_2 molecules. Therefore, the orbitals of this system are just the proper symmetry adapted linear combinations of the H_2 σ and σ^* orbitals and are shown in **2.16**. Furthermore, the a_1' and $1e'$ σ combinations along with the $2e'$ and a_2' σ^* combinations cannot be split very much in energy since the distances between the units, $r_2 = 4.0$ Å, is quite large. The orbitals of the cyclic H_6 molecule are given in reference 1 page 71 and are shown at the left hand side of **2.16**. On going from **a** to **b**, since r_1 decreases we expect the a_{1g} orbital of **a** to be stabilized, but r_2 increases which results in a destabilization. The two effects work in opposite directions and the net effect is cancellation, *i.e.*, the energy change of a_{1g} in **a** on moving to a_1' in **b** should be close to zero. A similar analysis can be used for the b_{1u} to a_2' conversion, *i.e.*, decreasing r_1

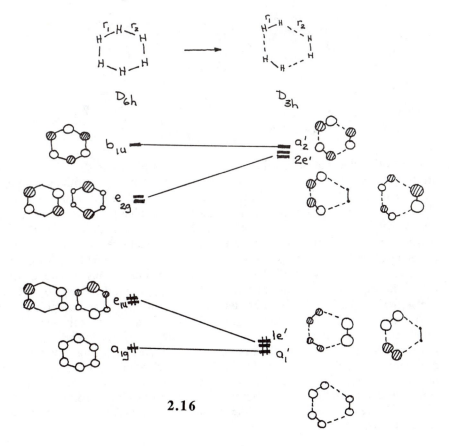

2.16

causes b_{1u} to rise in energy but increasing r_2 causes b_{1u} to be stabilized.

The change in energy of e_{1u} and e_{2g} upon distortion cannot be due to simple changes in overlap between the A.O.'s. Notice that the *form* of the orbitals are quite different at either side of the diagram. In **a** the H_6 molecule has D_{6h} symmetry, but when $r_1 \neq r_2$ the symmetry is reduced to D_{3h} and e_{1u} and e_{2g} both have e' symmetry. Consequently they can mix along the distortion path from **a** to **b**. This is a strongly avoided crossing (see p. 52-3 in reference 1) and results in a significant change in the degenerate wavefunctions such that $\psi(1e') \sim a\psi(e_{1u}) + b\psi(e_{2g})$ and $\psi(2e') \sim b\psi(e_{1u}) - a\psi(e_{2g})$ where a and b are mixing coefficients that depend upon how far we are along the distortion coordinate. The situation is shown in **2.17**.

$$\psi_{1e'} = a\left(\ \right) + b\left(\ \right) = \ $$

$$\psi'_{1e} = a\left(\ \right) + b\left(\ \right) = \ $$

$$\psi_{2e'} = b\left(\ \right) - a\left(\ \right) = \ $$

$$\psi'_{2e'} = b\left(\ \right) - a\left(\ \right) = \ $$

2.17

Notice that the phases in *both* components of e_{1u}, e_{2g}, $1e'$ and $2e'$ are arbitrary. Therefore, the precise way each component of e_{1u} mixes with e_{2g} is, in fact, dependent upon which way you have shaded e_{1u}, e_{2g}, etc. We have chosen (carefully and ahead of time) these specific phases, however $1e'$ could just as easily have been some $a\psi(e_{1u}) - b\psi(e_{2g})$ and $\psi(2e')$ or $-a\psi(e_{1u}) - b\psi(e_{2g})$ and $\psi(2e')$ their counterparts. The important result here is that an upper level will always stabilize a lower one when an avoided crossing takes place and a lower level will destabilize an upper one. Thus, whatever their actual description, e_{1u} becomes stabilized and e_{2g} destabilized on going to **b**. (See Question 2.7.) Since the four electrons in e_{1u} are stabilized on going from **a** to **b** then overall the distortion must be stabilizing. The solution of this problem has suprising generality. The nonexistence of cyclic Li_6, N_6 and P_6 have much in common with this problem. Since the orbital topology of hydrogen s orbitals is identical to p orbitals overlapping in a π-type fashion, the answer here offers some interesting insight as to whether aromaticity, *i.e.*, complete delocalization, is stabilizing or not. This question is addressed in 6.5. Finally it forecasts instability in linear chains with a particular electron count (Question 8.5.).

One of the problems in deciding on the energetics of processes like this one is the change in coordination number that occurs. Here the hydrogen atoms go from being two- to one-coordinate. The least of the problems is associated with balancing the stabilizing and destabilizing influences on an orbital as a result of the shortening and lengthening of interatomic separations. (A more serious problem is taking care of the often strong correlation effects which change on stretching chemical bonds.) If we assume that all of the short interatomic distance are the same in both **a** and **b**, then using simple Hückel theory the energies of the levels of cyclic H_6 are 2β, β(twice), $-\beta$(twice), -2β, and those of three H_2 units are β(three times) and $-\beta$(three times). This leads to a *destabilization* on going from **a** to **b** of 4β. One way of taking into account the variation in β with distance is to set the sum of the squares of the orbital energies equal as described in Question 2.10. (S. Lee, *Accts. Chem Res.*, **24**, 249 (1991)). This often mimics the change in β expected as the coordination number and bond lengths change. Thus if the energies of the levels of cyclic H_6 are 2β, β(twice), $-\beta$(twice), -2β, those of three H_2 units become $\sqrt{2}\beta$(three times) and $-\sqrt{2}\beta$(three times). This leads to a *stabilization* on going from **a** to **b** of 0.484β.

2.5.

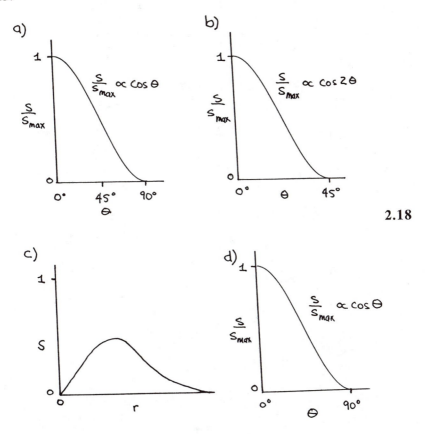

2.18

2.6.
(a)

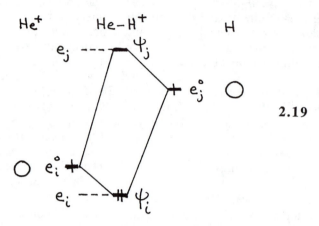

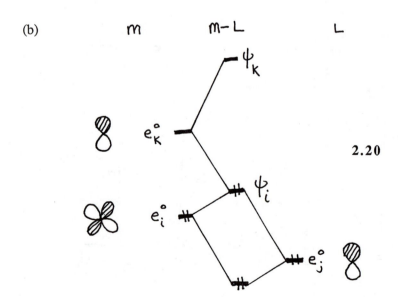

2.19

$e_i \propto e_i^0 + \tilde{S}_{ij}^2/(e_i^0 - e_j^0) = (+)^2/(-) = (-) \therefore$ stabilization
$e_j \propto e_j^0 + \tilde{S}_{ij}^2/(e_j^0 - e_i^0) = (+)^2/(+) = (+) \therefore$ destabilization
$\psi_i \approx \psi_i^0 + t_{ji}^{(1)}\psi_j^0, \quad t_{ji}^{(1)} \propto -\tilde{S}_{ij}/(e_i^0 - e_j^0) = (-)/(-) = (+)$

$\therefore \quad \psi_i = $ ◯→ $+ (—◯) = $ ◯—◯

$\psi_j \approx \psi_j^0 + t_{ij}^{(1)}\psi_i^0, \quad t_{ij}^{(1)} \propto -\tilde{S}_{ij}/(e_j^0 - e_i^0) = (-)/(+) = (-)$

$\therefore \quad \psi_j = $ —◯ $- ($◯—$) = $ ◐—◯

(b)

2.20

$e_j \propto e_j{}^0 + \tilde{S}_{ij}{}^2/(e_j{}^0 - e_i{}^0) + \tilde{S}_{jk}{}^2/(e_j{}^0 - e_k{}^0) = (+)^2/(-) + (+)^2/(-)$ ∴ stabilization,

$e_i \propto e_i{}^0 + \tilde{S}_{ij}{}^2/(e_i{}^0 - e_j{}^0) = (+)^2/(+) = (+)$ ∴ destabilization (Note that $\psi_k{}^0$ mixing in second order into $\psi_i{}^0$ actually causes a small third order energy change which is stabilizing.)

$e_k \propto e_k{}^0 + \tilde{S}_{jk}{}^2/(e_k{}^0 - e_j{}^0) = (+)^2/(+) = (+)$ ∴ destabilization (Note here the third order change is destabilizing.)

$\psi_j \approx \psi_j{}^0 + t_{ij}{}^{(1)}\psi_i{}^0 + t_{kj}{}^{(1)}\psi_k{}^0,\ t_{ij}{}^{(1)} \propto -\tilde{S}_{ij}/(e_j{}^0 - e_i{}^0) = (-)/(-) = (+)$
$\qquad\qquad t_{kj}{}^{(1)} \propto -\tilde{S}_{kj}/(e_j{}^0 - e_k{}^0) = (-)/(-) = (+)$

∴

$\psi_i \approx \psi_i{}^0 + t_{ji}{}^{(1)}\psi_j{}^0 + t_{ki}{}^{(2)}\psi_k{}^0,\ t_{ji}{}^{(1)} \propto -\tilde{S}_{ij}/(e_i{}^0 - e_j{}^0) = (-)/(+) = (-)$
$t_{ki}{}^{(2)} \propto \tilde{S}_{ij}\tilde{S}_{kj}/(e_i{}^0 - e_k{}^0)(e_i{}^0 - e_j{}^0) = (+)(+)/(-)(+) = (-)$

∴

$\psi_k \approx \psi_k{}^0 + t_{jk}{}^{(1)}\psi_j{}^0 + t_{ik}{}^{(2)}\psi_i{}^0,\ t_{jk}{}^{(1)} \propto -\tilde{S}_{jk}/(e_k{}^0 - e_j{}^0) = (-)/(+) = (-)$
$t_{ik}{}^{(2)} \propto \tilde{S}_{ij}\tilde{S}_{kj}/(e_k{}^0 - e_i{}^0)(e_k{}^0 - e_j{}^0) = (+)(+)/(+)(+) = (+).$

∴

(c)

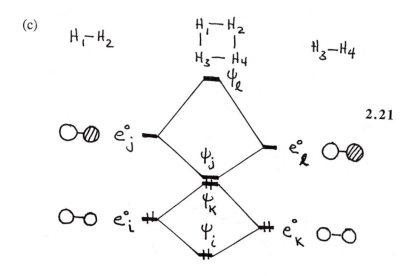

2.21

Note that since ,$<i/l> = <j/k> = 0$, all the non-degenerate interactions are zero and we only need to consider the degenerate ones. Thus

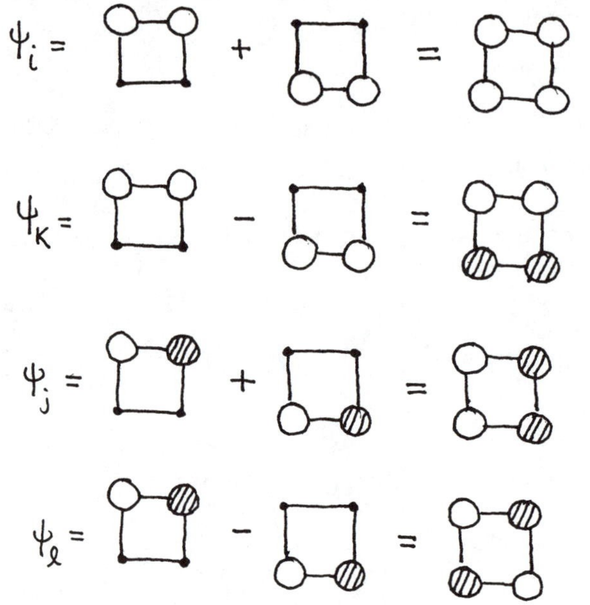

$$\psi_i \approx \psi_i^0 + \psi_k^0$$
$$\psi_k \approx \psi_i^0 - \psi_k^0$$
$$\psi_j \approx \psi_j^0 + \psi_l^0$$
$$\psi_l \approx \psi_j^0 - \psi_l^0$$

(d)

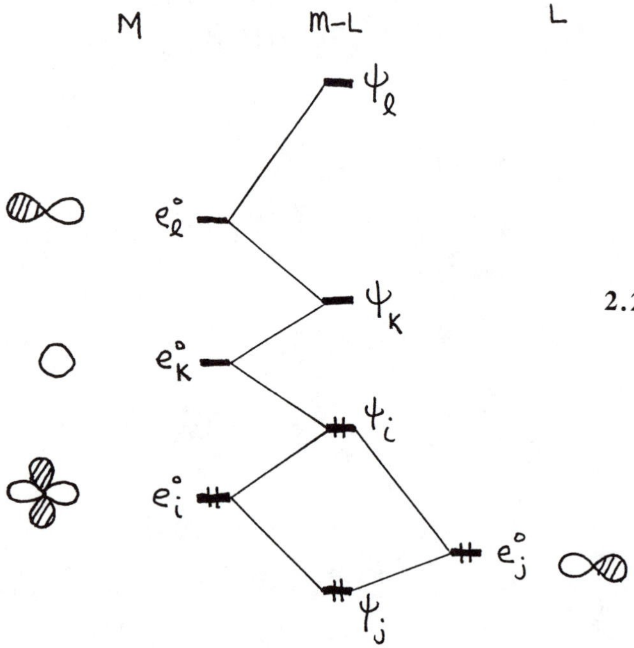

2.22

It should be an easy exercise by now to figure out whether or not each orbital is stabilized or destabilized. (**2.22** is drawn assuming that all $S_{ij} > 0$.)

$\psi_j \approx \psi_j^0 + t_{ij}^{(1)}\psi_i^0 + t_{kj}^{(1)}\psi_k^0 + t_{lk}^{(1)}\psi_l^0$, $t_{ij}^{(1)} \propto -\tilde{S}_{ij}/(e_j^0-e_i^0) = (+)$,
$t_{kj}^{(1)} \propto -\tilde{S}_{kj}/(e_j^0-e_k^0) = (+)$, $t_{lj}^{(1)} \propto -\tilde{S}_{lj}/(e_j^0-e_l^0) = (+)$. Here $t_{ij}^{(1)} > t_{kj}^{(1)} > t_{lk}^{(1)}$ because of the energy differences.

$\therefore$

$$\psi_j = \quad + \quad + \quad + \quad = $$

$\psi_i \approx \psi_i^0 + t_{ji}^{(1)}\psi_j^0 + t_{ki}^{(2)}\psi_k^0 + t_{li}^{(2)}\psi_l^0$, $t_{ji}^{(1)} \propto -\tilde{S}_{ji}/(e_i^0-e_j^0) = (-)$,
$t_{ki}^{(2)} \propto \tilde{S}_{ij}\tilde{S}_{jk}/(e_i^0-e_k^0)(e_i^0-e_j^0) = (-)$, $t_{li}^{(2)} \propto \tilde{S}_{ij}\tilde{S}_{jl}/(e_i^0-e_l^0)(e_i^0-e_j^0) = (-)$. Here $t_{ki}^{(2)} > t_{li}^{(2)}$ because of the energy differences.

$\therefore$

$$\psi_i = \quad - \quad - \quad - \quad = $$

$\psi_k \approx \psi_k^0 + t_{jk}^{(1)}\psi_j^0 + t_{ik}^{(2)}\psi_i^0 + t_{lk}^{(2)}\psi_l^0$, $t_{jk}^{(1)} \propto -\tilde{S}_{jk}/(e_k^0-e_j^0) = (-)$,
$t_{ik}^{(2)} \propto \tilde{S}_{ij}\tilde{S}_{kj}/(e_k^0-e_i^0)(e_k^0-e_j^0) = (+)$, $t_{lk}^{(2)} \propto \tilde{S}_{kj}\tilde{S}_{lj}/(e_k^0-e_l^0)(e_k^0-e_j^0) = (-)$. Here $t_{ik}^{(2)} \propto t_{lk}^{(2)}$.

$\therefore$

$$\psi_k = \quad - \quad + \quad - \quad = $$

$\psi_l \approx \psi_l^0 + t_{jl}^{(1)}\psi_j^0 + t_{il}^{(2)}\psi_i^0 + t_{kl}^{(2)}\psi_k^0$, $t_{jl}^{(1)} \propto -\tilde{S}_{jl}/(e_l^0-e_j^0) = (-)$,
$t_{il}^{(2)} \propto \tilde{S}_{ij}\tilde{S}_{lj}/(e_l^0-e_i^0)(e_l^0-e_j^0) = (+)$, $t_{kl}^{(2)} \propto \tilde{S}_{kj}\tilde{S}_{lj}/(e_l^0-e_k^0)(e_l^0-e_j^0) = (+)$. Here $t_{kl}^{(2)} > t_{il}^{(2)}$.

$\therefore$

$$\psi_l = \quad - \quad + \quad + \quad = $$

Notice that as a result of the orbital interactions, ψ_j is M-L bonding, ψ_l is M-L antibonding and ψ_i and ψ_k are basically nonbonding.

2.7.   The secular equations for this problem are just

$$\begin{vmatrix} H_{11} - E & H_{12} - S_{12}E \\ H_{12} - S_{12}E & H_{22} - E \end{vmatrix} \begin{pmatrix} c_1 \\ c_2 \end{pmatrix} = 0$$

The eigenvalues and eigenvectors may be obtained numerically by use of a suitable computer program or may be evaulated by hand as done now. By expansion of the secular determinant and solution of the resulting quadratic equation.

$$2E^{\pm} = (H_{11} + H_{22}) \pm \sqrt{[(H_{11} - H_{22})^2 + 4H_{12}^2]}$$

Now the first secular equation gives:

$$(H_{11} - E^{\pm})c_1^{\pm} + H_{12}c_2^{\pm} = 0$$

Since the wavefunctions must be normalized:

$$c_1^2 + c_2^2 = 1$$

So:

$$(c_1^{\pm})^2 = H_{12}^2/(H_{12}^2 + (H_{11} - E^{\pm})^2)$$
$$(c_2^{\pm})^2 = H_{12}^2/(H_{12}^2 + (H_{22} - E^{\pm})^2)$$

Thus for $H_{12} = 1$ we find the results in Table 2.8.

Table 2.8

x	$E^+(-E^-)$	$c_1^+(-c_2^-)$	$c_2^+(-c_1^-)$
0	5.099	0.0985	0.9951
1	4.123	0.1222	0.9925
2	3.162	0.1602	0.9871
3	2.236	0.2298	0.9732
4	1.414	0.3827	0.9239
5	1.000	0.7071	0.7071
6	1.414	0.9739	0.3827
7	2.296	0.9732	0.2298
8	3.162	0.9871	0.1602
9	4.123	0.9925	0.1222
10	5.099	0.9951	0.0985

These results are plotted graphically in **2.23**. Notice the avoided crossing at x = 5 and the mixing between the two functions which gets larger the closer to this point. The mixing betwen the two is clearly controlled by the size of H_{12}. Numerically when $H_{12} = 0.01$ we find the figures in Table 2.9.

Table 2.9

x	$E^+(-E^-)$	$c_1^+(-c_2^-)$	$c_2^+(-c_1^-)$
4	1.00	0.0050	1.0000
5	0.01	0.7071	0.7071
6	1.00	1.0000	0.0050

which shows a significant energy shift and orbital mixing only at $x = 5$.

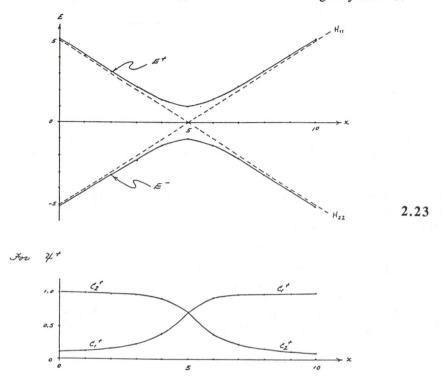

2.8. Notice that it is the ring carbon resonances which are shifted in frequency. These are the atoms which carry the $p\pi$ orbitals which are the major players in the interaction of the ligands with the metal. From **5.45** we can see that these ligand orbitals are involved both in the e_{1g} and e_{2g} orbitals of the metallocene, the *HOMO* and *LUMO* of the closed-shell molecule ferrocene. If the manganocene is low-spin (as in *(EtMe4Cp)2Mn*) then the one unpaired electron occupies the e_{2g} orbital. If the manganocene is high-spin (as in *(MeCp)2Mn*) with five unpaired electrons, then there is one electron in each component of e_{1g} and e_{2g}, and one electron in a_{1g}. All three orbitals are of π type, located on the ring carbon atoms with little or no contribution from the carbon atoms of the methyl or ethyl groups.

2.9.   The point groups of CH_4 and CH_3D are T_d and C_{3v} respectively. The molecular orbital diagram for CH_4 is shown in **2.30** with the correct symmetry labels for the T_d point group. We are thus interested in the way the a_1 and t_2 labels change as the symmetry is lowered on substitution. Table 2.10 shows the characters for these two symmetry species extracted from the relevant character table of Chapter I. We need to correlate these characters with the symmetry elements that are retained as the symmetry drops.
A_1 (of T_d) clearly becomes A_1 (in C_{3v}), but the representation T_2 (of T_d) is reducible; no three-dimensional representations being allowed in this point

Table 2.10.

T_d	E	$8C_3$	$3C_2$	$6S_4$	$6\sigma_v$
A_1	1	1	1	1	1
T_2	3	0	-1	-1	1

C_{3v}	E	$2C_3$	$3\sigma_v$
A_1(of T_d)	1	1	1
T_2(of T_d)	3	0	1

group. Inspection of the character table shows it to be $A_1 + E$ (in C_{3v}). Thus $a_1 \rightarrow a_1$ and $t_2 \rightarrow a_1 + e$. Such a lowering of symmetry shows up most strongly in the vibrational spectrum, rather than in the electronic properties since the *C-H* and *C-D* distances (and thus the relevant overlap integrals) are very similar indeed.

2.10. **2.24** shows the calculated energy levels (in units of β) for the two structures. The figures in parentheses are obtained by setting the second moments of both structures equal. You can readily show that the second moment for each structure is simply equal to *number of H-H contacts x 2β*. Thus the energies of the *spiro* structure are scaled by $\sqrt{(18/12)} = \sqrt{(3/2)}$. It is

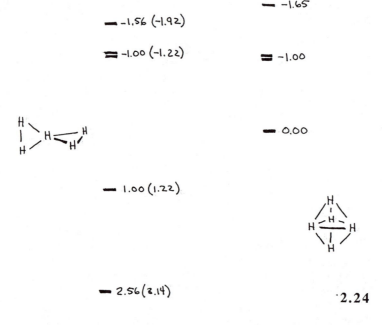

now easy to calculate that H_5^{3+} prefers the trigonal bipyramidal structure (7.3β) over the *spiro* one (6.3β), but that H_5^+ prefers the *spiro* structure (8.7β) over the trigonal bipyramidal one (7.3β).

2.11. Determination of the symmetry species of the singlet and triplet states of the atom proceeds along exactly the same lines used for molecules (*e.g.*, Question 6.1.) and requires evaluation of the symmetric and antisymmetric direct products whenever a degenerate level is filled with two electrons.
(a) This is a closed shell and thus leads to a ^1S state.
(b) Here the symmetric and antisymmetric direct products of p are required and are evaluated in Table 2.11. The result is $^1S + {}^1D + {}^3P$.

Table 2.11.

for p	E	C_4	C_2	C_3	
$\chi(\mathcal{R})$	3	1	-1	0	
$\chi(\mathcal{R}^2)$	3	-1	3	0	
$\chi^2(\mathcal{R})$	9	1	1	0	
$\chi^2_{sym}(\mathcal{R})$	6	0	2	0	$(s+d)$
$\chi^2_{antisym}(\mathcal{R})$	3	1	-1	0	(p)

(c) Here the symmetric and antisymmetric direct products of d are required and are evaluated in Table 2.12. The result is $^1S + {}^1D + {}^1G + {}^3F + {}^3P$.

Table 2.12.

for d	E	C_4	C_2	C_3	
$\chi(\mathcal{R})$	5	-1	1	-1	
$\chi(\mathcal{R}^2)$	5	1	5	-1	
$\chi^2(\mathcal{R})$	25	1	1	1	
$\chi^2_{sym}(\mathcal{R})$	15	1	3	0	$(s+d+g)$
$\chi^2_{antisym}(\mathcal{R})$	10	0	-2	1	$(p+f)$

(d) This is just a case with two holes rather than two electrons. The symmetry species of the electronic states are the same as for (c).

2.12. The starting orbitals are given in **2.25**, the oxygen orbitals lying more deeply than the metal ones. The orbital phases have been set so as to give rise to a positive overlap between the orbitals on the two centers. This will simplify the perturbation treatment below. The orbitals of δ type remain nonbonding since there are no ligand orbitals of this symmetry, and the π orbitals interact to give bonding and antibonding pairs. The level shifts in **2.25** and the form of the wavefunctions shown in **2.26** are easy to understand for the π-type

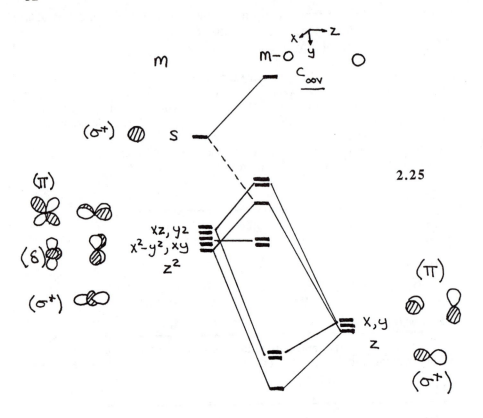

2.25

interactions. For example:

$$e_{xz} \propto e_{xz}^0 + S_{y,yx}^2/(e_{yz}^0 - e_y^0) = e_{xz}^0 + (+)/(+), \text{ i.e., destabilized.}$$

and $\psi_{xz} \approx \psi_{xz}^0 + t_{y,yx}^{(1)}\psi_y \propto \psi_{xz}^0 - S_{y,yx}/(e_y^0 - e_{yz}^0).\psi_y = \psi_{xz}^0 - (-)(+)/(+)\psi_y$

as shown in **2.26**. The energetics and orbital description of the σ type orbitals are a little more complex. Writing z, z^2 and s as $\psi_{1\text{-}3}$,

$$e_1 \propto e_1^0 + S_{1,2}^2/(e_1^0 - e_2^0) + S_{1,3}^2/(e_1^0 - e_3^0) = e_1^0 + (+)/(-) + (+)/(-), \text{ i.e.,}$$
stabilized.

$e_2 \propto e_2^0 + S_{1,2}^2/(e_2^0 - e_1^0) + \text{(third order term involving s)} = e_2^0 + (+)/(+) + \text{(small stabilization)}, \text{ i.e., destabilized.}$

$$e_3 \propto e_3^0 + S_{1,3}^2/(e_3^0 - e_1^0) = e_3^0 + (+)/(+), \text{ i.e., destabilized.}$$

and

$$\psi_1 \approx \psi_1^0 + t_{2,1}^{(1)}\psi_2 + t_{3,1}^{(1)}\psi_3$$
$$\propto \psi_1^0 - S_{1,2}/(e_1^0 - e_2^0).\psi_2 - S_{1,3}/(e_1^0 - e_3^0).\psi_3$$
$$= \psi_1^0 + (-)(+)/(-).\psi_2 + (-)(+)/(-).\psi_3$$

$$\psi_2 \approx \psi_2^0 + t_{1,2}^{(1)}\psi_2 + t_{3,2}^{(2)}\psi_3$$
$$\propto \psi_2^0 - S_{1,2}/(e_2^0 - e_1^0).\psi_2 - S_{1,3}S_{2,3}/(e_2^0 - e_3^0)/(e_2^0 - e_1^0).\psi_3$$
$$= \psi_2^0 + (-)(+)/(+).\psi_2 + (+)(+)/(-)(+).\psi_3$$

$$\psi_3 \approx \psi_3^0 + t_{1,3}^{(1)}\psi_1 + t_{2,3}^{(2)}\psi_2$$
$$\propto \psi_3^0 - S_{1,3}/(e_3^0 - e_1^0).\psi_1 - S_{1,2}S_{1,3}/(e_3^0 - e_2^0)/(e_3^0 - e_1^0).\psi_2$$
$$= \psi_3^0 + (-)(+)/(+).\psi_2 + (+)(+)/(+)(+).\psi_2$$

2.13. We start off by constructing the orbitals of triangular H_3. These are simple to derive using group theory. The three $1s$ orbitals transform as $a_1' + e'$ and the relevant symmetry adapted combinations of these orbitals are shown in **2.27**. The energies of the molecular orbitals are evaluated by determining the

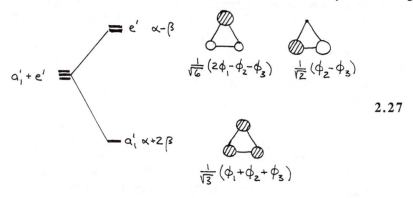

$$e' \quad \alpha - \beta$$

$$\frac{1}{\sqrt{6}}(2\phi_1 - \phi_2 - \phi_3) \qquad \frac{1}{\sqrt{2}}(\phi_2 - \phi_3)$$

$$a_1' + e'$$

2.27

$$a_1' \quad \alpha + 2\beta$$

$$\frac{1}{\sqrt{3}}(\phi_1 + \phi_2 + \phi_3)$$

value of $<i/\mathcal{H}eff/i>$. Within the Hückel approximation $e(a_1') = \alpha + 2\beta$, and $e(e')$ $= \alpha - \beta$. The electronic configurations of H_3^+, H_3 and H_3^- are simply $(a_1')^2$, $(a_1')^2(e')^1$ and $(a_1')^2(e')^2$ respectively. The symmetry species of the resulting electronic states for the first two species are readily written down as $^1A_1'$ and $^2E'$ respectively. Those for H_3^- are obtained by evaluating the symmetric and antisymmetric direct products of e'. This is shown in Table 2.13.

Table 2.13.

for e'	E	$2C_3$	$3C_2$	σ_h	$2S_3$	$3\sigma_h$	
$\chi(\mathcal{R})$	2	-1	0	2	-1	0	
$\chi(\mathcal{R}^2)$	2	-1	2	2	-1	2	
$\chi^2(\mathcal{R})$	4	1	0	4	1	0	
$\chi^2_{sym}(\mathcal{R})$	3	0	1	3	0	1	$(= a_1' + e')$
$\chi^2_{antisym}(\mathcal{R})$	1	1	-1	1	1	-1	$(= a_2')$

Thus the electronic states for H_3^- are $^1A_1' + {}^1E' + {}^3A_2'$. We may then draw the following conclusions concerning the structures of these molecules. H_3^+ should be stable as an equilateral triangle $(^1A_1')$ which it is. H_3 $(^2E')$ should be Jahn-Teller unstable and distort away from this geometry. H_3^-, if it is a triplet $(^3A_2')$ should be stable in this geometry, but as a singlet $(^1E'$ will probably lie lower in energy) will be Jahn-Teller unstable. The symmetry species of the Jahn-Teller active distortion mode is given by the symmetric direct product of e', which as we have shown is $a_1' + e'$. The two components of the e' vibration are shown in **2.28** and take the equilateral triangular structure either to a

a   **2.28**   b

isoceles triangular or linear geometry.
    2.29 shows assembly of the molecular orbital diagram for linear H_3 using the ideas of group theory. The σ_u^+ orbital is nonbonding with an energy of α and the pair of σ_g^+ orbitals have energies of $\alpha \pm \sqrt{2}\beta$. Thus H_3^+ is more stable in the triangular $(2\alpha + 4\beta)$ than linear $(2\alpha + 2\sqrt{2}\beta)$ geometry, but H_3^- more stable in the linear geometry $(4\alpha + 2\sqrt{2}\beta)$ than in the triangular geometry $(4\alpha + 2\beta)$.

2.14. (a) The carbon $2s$ and $2p$ orbitals transform as $a_1 + t_2$ respectively. The hydrogen $1s$ orbitals transform as $a_1 + t_2$ too. Assembly of the molecular orbital diagram is thus straightforward and is shown in **2.30**. Bonding and antibonding orbitals of each symmetry result. The energetic location of the starting atomic orbitals is determined by their ionization energies. Numerical

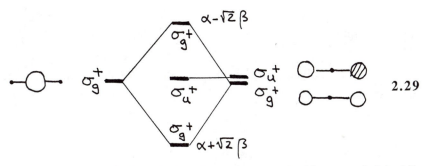

2.29

values are given in Table 1.2 (b) There are four doubly occupied bonding orbitals and four bond pairs in the Lewis picture. Notice that although the bonding character associated with each *C-H* linkage is identical, the four bonding pairs of electrons are not equivalent on the molecular orbital picture, there being one involving $s(a_1)$ and three involving carbon $p(t_2)$ orbitals. There is clearly no *sp* hybridization in the molecule. Such mixing of *s* and *p* orbitals is forbidden by symmetry. (c) The photoelectron spectrum of the molecule shows ionization from a deep lying orbital close to where we expect to find a carbon *2s* orbital (the *1a₁* molecular orbital) and ionization from levels close to where we expect to find a carbon *2p* orbital (the *1t₂* molecular

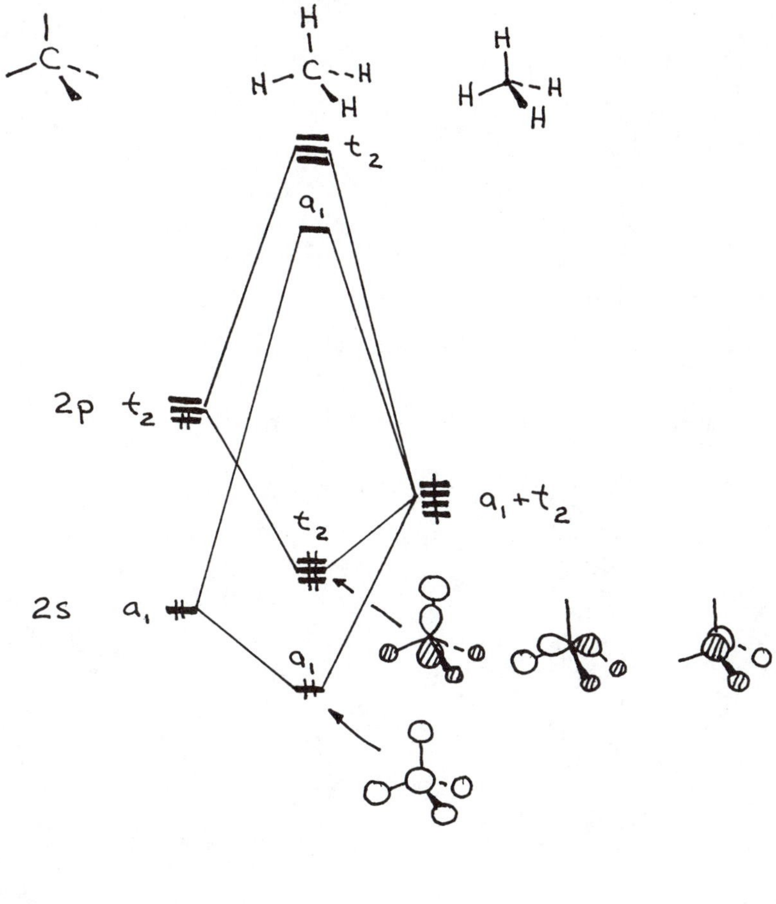

2.30

orbitals) and is in accord with our molecular orbital picture. A model which contained four equivalent bonding orbitals based on four equivalent sp^3 hybrids would not leads to such a picture.

We can however generate the traditional hybrid picture by taking suitable linear combinations of the $1a_1$ and $1t_2$ molecular orbitals as shown in **2.31**. As discussed in the answer to Question 2.3. these are not stationary states

$$\hspace{6cm} 2.31$$

of the Schrödinger equation, but are very useful nevertheless. Experimental work using *e-2e* spectroscopy has revealed the orbital composition of these occupied orbitals. They are indeed just as shown in **2.30**.

2.15. (a) Label the hydrogen $1s$ orbital as χ_1 and the lithium $2s$ orbital as χ_2. Then, $\langle\chi_1|\chi_2\rangle = S_{12} = 0.3609$, $H_{11} = e_1^0 = -13.60\text{eV}$, $H_{22} = e_2^0 = -5.40\text{eV}$, and using the Wolfsberg-Helmholtz approximation $H_{12} = \Delta_{12} = K(H_{11} + H_{22})S_{12}/2 = -6.00\text{eV}$. Thus the secular determinant is

$$\begin{vmatrix} e_1^0 - e & \Delta_{12} - S_{12}e \\ \Delta_{12} - S_{12}e & e_2^0 - e \end{vmatrix} = 0$$

with roots $e_\pm = (-b \pm \sqrt{D})/2a$, where

$$a = 1 - S_{12}^2 = 0.8698$$
$$b = 2\Delta_{12}S_{12} - e_1^0 - e_2^0 = 14.6692\text{eV}$$
$$D = b^2 - 4ac \text{ with } c = e_1^0e_2^0 - \Delta_{12}^2 = 37.44\text{eV}^2$$
so $\quad D = 84.9242\text{eV}^2$

and $\quad e_1 = -13.73\text{eV}$, $e_2 = -3.14\text{eV}$.

The two wavefunctions are of the form $\psi_i = \Sigma_j c_{ij}\chi_i$ with

$$c_{11} = 1/\sqrt{(1 + 2tS_{12} + t^2)} \text{ with } t = \Delta_{12} - S_{12}e_1^0/(e_1^0 - e_2^0) = 0.1331$$

so $\quad c_{11} = 0.9475.$

$$c_{21} = tc_{11} = 0.1261$$

Similarly, $\quad c_{22} = 1/\sqrt{(1 + 2t'S_{12} + t'^2)}$ with $t' = \Delta_{12} - e_2^0S_{12}/(e_1^0 - e_2^0)$
$$= -0.4940$$

so $\quad c_{22} = 1.0615.$

$$c_{12} = t'c_{22} = -0.5244.$$

(b) The Mulliken populations are simply $P_{ij} = (2-\delta_{ij})\Sigma_k n_k S_{ij} c_{ik} c_{jk}$ where n_k is the number of electrons in the kth molecular orbital and $\delta_{ij} = 1$ if $i = j$ and 0 if $i \neq j$. For this problem $n_1 = 2$ and $n_2 = 0$. Thus:

$$P_{11} = (2)(0.9475)^2 = 1.7955$$
$$P_{22} = (2)(0.1261)^2 = 0.0318$$
$$P_{12} = (2)(2)(0.9475)(0.1261)(0.3609) = 0.1725.$$

Dividing the overlap population equally between the two atoms leads to $1.7955 + (0.1725)/2 = 1.8818$ electrons on hydrogen (*i.e.*, a charge of -0.8818) and $0.03318 + (0.1725)/2 = 0.1182$ electrons on lithium (*i.e.*, a charge of +0.8818). This is ignoring the two electrons in the $1s$ shell of lithium. Thus $\mu^* = 2.54(0.8818)(3.015) = 6.75$ Debye.

(c) The primary reason why our calculation above is different from that found at the *ab initio* level is that the $2p$ orbitals have been neglected on lithium. Their effect is shown in **2.32**. Notice that the inclusion of the lithium p orbitals leads to admixture into ψ_1 leading to an increase in electron density on Li and a decrease in the charge separation. For the same reason Q_H becomes less negative and P_{LiH} increases. In the process more covalent character is introduced into the Li-H bond and the value of μ^* decreases. On going from the minimal to extended basis levels, the extra functions serve to increase the ionic character, the charges become larger, P_{LiH} decreases and μ^* increases.

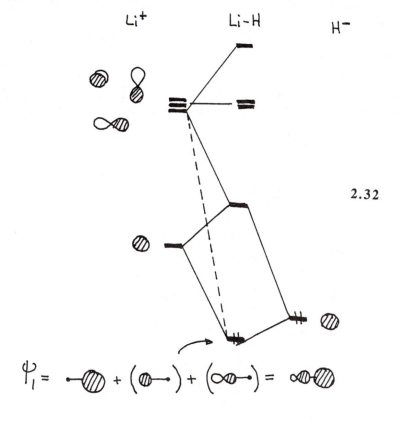

(d) The dipole moment is really the gradient of the charge density over all space. In a diatomic molecule one might normally assume that most of the charge lies between the nuclei and hence can be evaluated in terms of point charges centered at the nuclei. However, there is in this case considerable mixing of the lithium $2p$ atomic orbital into ψ_1. Consequently some electron density extends beyond lithium and away from hydrogen. The effective distance associated with a part of this charge density has thus increased dramatically. The reason why E_1 lies so much higher in energy in the *ab initio* calculation is due to electron-electron repulsions, formally ignored in the one-electron calculation. Increasing the flexibility of the basis set allows the electron density to spread out to some extent so that in the extended basis E_1 is lowered.

2.16. (a) For each member of the *HA* series there are always two electrons in the deepest-lying *HA* σ bonding orbital leading to a σ bond. All additional electrons, 1 in *HBe* up to 6 in *HF*, enter *nonbonding* orbitals as indicated by the orbital diagram of **4.43**, a result suggested too by Lewis arguments. What does change is the electronegativity of the atom *A*. As it becomes more electronegative the atomic orbitals contract and the *H-A* bond length must decrease to maintain overlap with the *H 1s* orbital. Another way of expressing the same idea is to say that the 'size' of the atom is becoming smaller on moving to the right of the periodic table. The dissociation energy, *D.E.*, (recall this is the energy of *HA* relative to $H\cdot + A\cdot$) increases because the energy of the electron donated by hydrogen, now increasingly located on *A* lies in an orbital which is getting deeper in energy.

(b) **2.33** shows the molecular orbitals of an A_2 diatomic molecule. Clearly $2\sigma_u^+$ and π_g are strongly antibonding, whereas, $1\sigma_g^+$ and π_u are strongly bonding. While $1\sigma_u^+$ is in all likelihood nonbonding, it's not so clear whether $2\sigma_g^+$ should be a bonding or nonbonding orbital. In fact numerical calculations show that the nature of both of these orbitals is quite sensitive to the identity of

$2\sigma_u^+$

π_g

$2\sigma_g^+$ 2.33

π_u

$1\sigma_u^+$

$1\sigma_g^+$

the atoms concerned. Data for the series is shown in Table 2.14.

Table 2.14

molecule	electron configuration	length (Å)	D.E.(kcal/mol)
C_2	$(1\sigma_g{}^+)^2(1\sigma_u{}^+)^2(\pi_u)^4$	1.24	144
CN	$(1\sigma_g{}^+)^2(1\sigma_u{}^+)^2(\pi_u)^4(2\sigma_g{}^+)^1$	1.17	188
CO	$(1\sigma_g{}^+)^2(1\sigma_u{}^+)^2(\pi_u)^4(2\sigma_g{}^+)^2$	1.13	256
CF	$(1\sigma_g{}^+)^2(1\sigma_u{}^+)^2(\pi_u)^4(2\sigma_g{}^+)^2$ $(\pi_g)^1$	1.27	131

We could argue that going from C_2 to CO is just like the HA series in (a) and that $2\sigma_g{}^+$ is nonbonding. The change in D.E. is, however, a good bit larger than that for HC to HO. Perhaps then $2\sigma_g{}^+$ is slightly bonding. Clearly π_g is antibonding and population of this orbital in CF will result in lengthening the bond and lead to a smaller dissociation energy.

(c) Here all of the molecules are isoelectronic in terms of the number of valence electrons. Because the radial maxima of the np atomic orbitals increase in the order $2p<3p<4p<5p$ one expects bond lengths to surrounding atoms to increase with increasing n. Likewise, the overlaps in the σ and particularly in the π sense go as $S_{2p,2p} > S_{3p,2p} > S_{3p,3p} \sim S_{4p,2p} > S_{4p,3p}$. The order of the bond lengths are then easy to understand. Notice that the bond lengths and dissociation energies of CS and SiO are very close to each other in accord with this idea. SiS and SnO should also be similar on this simple model. However, the bond length in the latter is the shorter of the two but the dissociation energy is the larger.

2.17. For H_2 and $H_2{}^+$ the occupied bonding orbital is of the form $\psi = (1/\sqrt{2})(\phi_1 + \phi_2)$ where the ϕ_i are the hydrogen $1s$ orbitals. In triangular $H_3{}^+$ the occupied bonding orbital is of the form $\psi = (1/\sqrt{3})(\phi_1 + \phi_2 + \phi_3)$. Overlap has been ignored in the normalization. The bond overlap populations are thus given as in Table 2.15 where S is the overlap integral $<\phi_1/\phi_2>$.

Table 2.15

Molecule	Overlap Population	Bond Length (Å)
H_2	2S	0.74
$H_3{}^+$	(4/3)S	0.87
$H_2{}^+$	S	1.06

By extrapolation between the values for H_2 and $H_2{}^+$ a bond length of around 0.95Å might be expected for $H_3{}^+$. In fact it is somewhat shorter at 0.87Å.

2.18. The first step is to construct the orbitals of H_4 in this geometry. This is easy to do using the orbitals of triangular H_3 and is shown in **2.34**. It will be particularly important here to get the form of the orbitals correct. They are

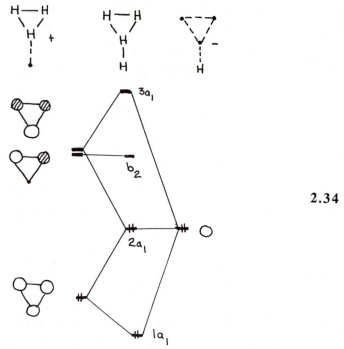

2.34

shown in **2.35**. For an electronegativity perturbation $e_i(1) = (C_{\alpha i}^0)^2 \delta\alpha$, $e_i(2) = \sum_{j \neq i} (C_{\alpha i}^0 C_{\alpha j}^0)(\delta\alpha)^2/(e_i^0 - e_j^0)$ with $|e_i(1)| > e_i(2)$, and $\delta\alpha < 0$ for *He* but > 0 for *Be*. The results are shown in **2.36** and **2.37**. The important point to note is the good stabilization of $2a_1$ in case 1 for *He* substitution which is absent for case

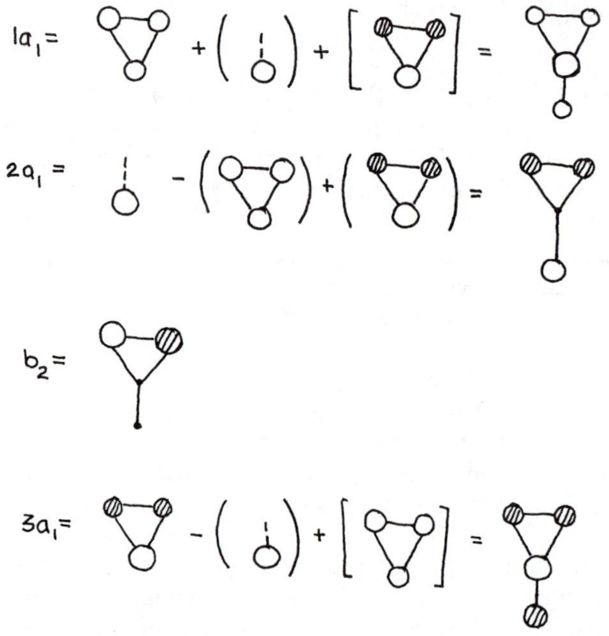

2.35

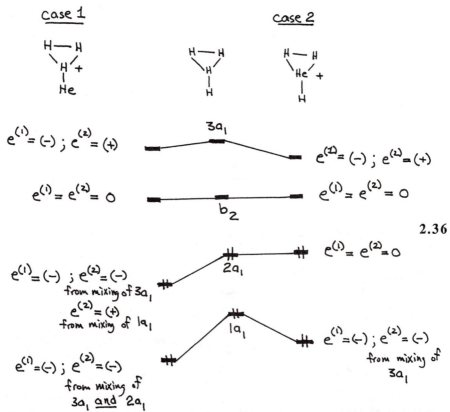

2. This comes about because there is a node at the ring site in this orbital. The stability of case 1 over case 2 is derived from this difference. For Be substitution it is the lack of movement of this orbital because of its nodal properties which makes case 2 stable here.

2.19. (a) **2.38** shows the orbital character from the calculation. Most of the orbitals are simply derived from those of the homoatomic molecule *via* first-order perturbation theory. Since oxygen is more electronegative than carbon, the π bonding levels are largely O located and the π antibonding levels largely C located. The same is true of the σ bonding and antibonding orbitals. The situation is a little more complex for the σ nonbonding orbitals ψ_4 and ψ_7 and their description results from mixing with other orbitals of σ symmetry.

(b) It is important to realize that the formal charge assigned to an atom is a somewhat arbitrary way of apportioning electrons. An easily understood case in point would be H_4N^+. *Any* calculation when the atomic charges are summed will add up to +1.000. However, a +1.00 charge at N implies that of the eight valence electrons, four reside at N. For this to be true, the coefficients at N would have to be the same as those on H for the four occupied levels and this means, in turn, that the electronegativity of N must be about the same as H. Clearly this is not true. N is more electronegative than H and, thus, one expects larger coefficients on N for the occupied levels. This, of course, then leads to a larger charge density at N and, in fact, it can easily have a negative

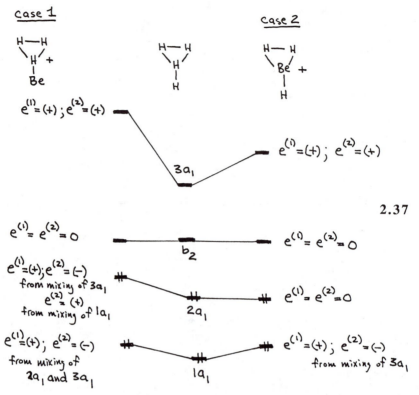

charge computed for it. The hydrogens, being *less* electronegative than *N* will have a positive charge associated with them.

For the *CO* example, although one can write it as $\cdot:C\equiv O:^+$, the same considerations apply. Most of the bonding levels will be concentrated on *O* and, therefore, it will be expected to have a negative charge while *C* has an equal positive charge.

The variational theorem tells us that one can only calculate an upper limit for the total energy. Thus, the lower the energy is calculated to be, the closer it is to reality. The same is true for the wave function. Thus, the calculated energy drops greatly going from single zeta to a double zeta basis and then less so for the third basis set and finally very little energy gain (3kcal/mole!) is observed on going to the fourth basis set. What is clear, however, is that the computed charges vary greatly going from one basis set to another. The Mulliken population analysis is *not* "variational". Adding *f* functions certainly does not change the bonding in *CO*, otherwise the energy would drop accordingly. However, the charges on *C* and *O* do change. What happens is that very small amounts of *f* orbital character mix into the occupied levels. This causes the overlap population to change. While this mixing is small, the number of *f* functions is large and, therefore, the total effect is sizeable. In the Mulliken population analysis the overlap population is divided by half and assigned to each atom. This is an entirely arbitrary assignment of the electron density shared between two atoms. There are more sophisticated (but still just as arbitrary) ways of partitioning the electron density. However,

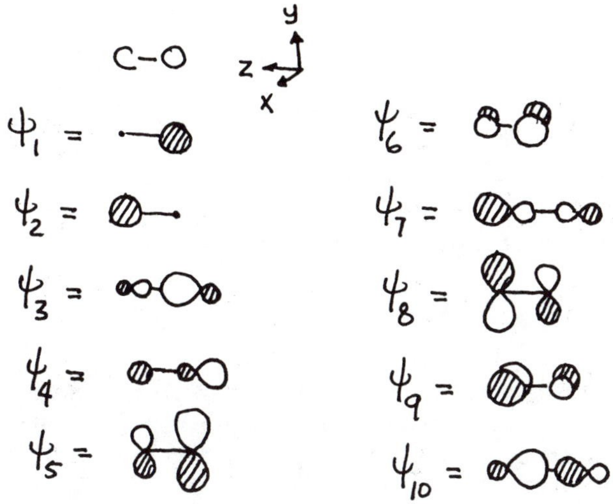

$$C-O$$

$$\psi_1 = \qquad \qquad \psi_6 =$$

$$\psi_2 = \qquad \qquad \psi_7 =$$

$$\psi_3 = \qquad \qquad \psi_8 = \qquad 2.38$$

$$\psi_4 = \qquad \qquad \psi_9 =$$

$$\psi_5 = \qquad \qquad \psi_{10} =$$

we do this means that the computed charges will necessarily also vary a sizable amount as the basis set is changed. Thus, if the exponents for the *f* functions were changed or *g* functions were added to the basis set, the change in the total energy would probably be extremely small, however, there still would be considerable changes in the computed charges. These results emphasize the fact that the details of the electronic charge distributions are much more sensitive to the nature of the calculation than the geometry. Thus crude models are often good enough to get the details of molecular geometries correct, but a much higher quality calculation is needed to get dipole moments and quadrupoles correct.

Chapter III.

Transition Metal Chemistry

3.1. Use the angular overlap model to show that $\Delta_{tet} = 4/9\Delta_{oct}$ for octahedral and tetrahedral transition metal complexes. What assumptions did you make in your derivation? Are low spin tetrahedral complexes expected to be prevalent? Using the AOM diagrams for octahedral, square pyramidal and square complexes, calculate the $MOSE$ values for the two appropriate d^8 configurations. Can you use the result to understand the observation that for many Ni^{II} compounds there seems to be an equilibrium between four-, five-, and six-coordinated species in solution.

3.2. Explain the observation that whereas $Ni^{II}(CN)_5^{3-}$ has a square pyramidal geometry with Ni-C distances of 2.17Å (axial) and 1.85Å (basal), $Ni^{II}(5$-$ClsalenNEt_2)_2$ has Ni-N bond lengths of 1.98Å (axial) and 2.00Å (basal).

3.3. Assemble a molecular orbital diagram from the appropriate fragments for a symmetric X_5M-Y-$M'X_5$ species (3.1) and hence show why such molecules are not known as stable entities for $M, M' = Cr^{II}$ and Cr^{III}, and $X = H_2O, Y =$ halide, nor for similar species involving Pt^{II} and Pt^{IV}. Such species are in fact probably the transition states for inner-sphere electron transfer reactions.

3.1

3.4. Derive the molecular orbital diagram for a transition metal ML_6 complex by examining the symmetry properties of the ligand σ and π orbitals and the central atom s, p and d orbitals. Use your results to explain the following. (a) $Fe(CN)_6^{4-}$ has a shorter Fe-C distance than $Fe(CN)_6^{3-}$ even though the ionic radius of Fe^{2+} is larger than that of Fe^{3+}. (b) The infra-red spectrum of MCl_6^{-3} complexes show a single band in the M-Cl stretching region at 315cm^{-1} for $M = Cr$ and at 248cm^{-1} for $M = Fe$. What would you predict for $M = Mn$?

3.5. (a) Construct a molecular orbital diagram for a square pyramidal MH_5 molecule. (b) Focus on the d orbitals alone and construct a Walsh diagram associated with changing the apical-basal angle. (c) Use your result to understand the (biologically vital) movement of the iron atom in Hemoglobin

3.2

on oxygen coordination (**3.2**). On coordination the iron atom changes from high spin to low spin. Are there any other factors not included in your model? The details of the angular dependence of metal d-ligand σ overlap integrals are given in Table 1.4. (d) Show why $Cr(CO)_5$ has a square pyramidal structure but $Fe(CO)_5$ a trigonal bipyramidal one with a square pyramidal structure a little higher in energy.

3.6. Adamson's rule, which may be used to predict the pathway of ligand dissociation on photochemical exitation of Werner-type transition metal complexes, may be phrased as follows. The ligand that is lost is that with the stronger 'ligand field' (Δ) which lies along the weaker field axis (x, y, or z) of the complex. By studying how the e_g energy levels of the transition metal complex split apart in energy as the molecule with O_h symmetry is substituted by different ligands, provide an orbital rationalization for this rule. (As a general rule of thumb photodissociation usually occurs from the lowest lying excited state of the molecule.)

3.7. Two different geometries have been established for low-spin d^6 five coordinate molecules, the first (**a**) with $\alpha \sim 80°$ and the second (**c**) with $\alpha \sim 180°$ (**3.3**). The latter is the square pyramid found for $Cr(CO)_5$ and the former the structure elucidated for $Ir(PR_3)_2HClPh$. Show how in an ML_4X

3.3

complex the first structure is favored when X is a π-donor and the second when it is a π-acceptor. What are the criteria for the observation of a geometry with $\alpha \sim 120°$?

3.8. The CO stretching frequencies of the isoelectronic molecules $V(CO)_6^-$, $Cr(CO)_6$ and $Mn(CO)_6^+$ increase in this order. Explain this observation in molecular orbital terms.

3.9. The molecule $Cr(CO)_5$ undergoes a photochemically induced rearrangement process which may be detected either by experiments on CS

substituted systems or by using polarized photolysis. The effect is to exchange axial and basal linkages. By constructing an orbital diagram which correlates the levels of the square pyramid with those of the trigonal bipyramid, show how such a photochemically induced Berry process may occur which would lead to this result.

3.10. Derive a molecular orbital diagram for a tri-capped trigonal prismatic MH_9 molecule (**3.4**). (ReH_9^{2-} actually has this structure.) Start off by

                                                                      3.4

generating the correct symmetry adapted linear combinations of ligand orbitals. Construct proper linear combinations of orbitals of the same symmetry but which come from symmetry inequivalent sets, and finally assemble the complete diagram.

3.11. Construct an orbital correlation diagram linking octahedral and square planar ML_n species via the loss of two *trans* ligands. Now construct a similar diagram but using the lowest electronic *states* of the d^8 electron configuration. Use your results to provide a qualitative explanation of the observation that whereas $Ni(H_2O)_6^{+2}$ is a paramagnetic octahedral molecule, $Pt(CN)_4^{-2}$ is a diamagnetic square planar molecule.

3.12. Predict the structure of the species $Rh(PPh_3)_3^+$.

3.13. For the trigonal bipyramidal ML_5 geometry, which should be shorter for a low-spin d^8 transition metal complex, the axial or equatorial M-L linkages?

3.14. Square pyramidal $Cr(CO)_5$ can be made with an ostensibly inert species (X) coordinated to the sixth site in 'inert gas' matrices. However, **3.5** shows the dramatic shift of the lowest energy d-d electronic transition as a function of matrix medium. (a) Construct a molecular orbital diagram for the six

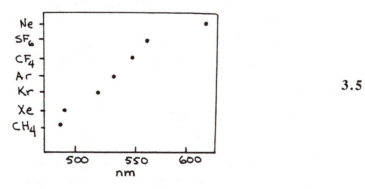

                                                                      3.5

coordinate molecule showing the effect of a sixth ligand on the orbital energies. (b) Create a model to rationalize the data in the figure. (c) Use the ionization potentials of the sixth site occupants and perturbation theory to numerically correlate the strength of interaction with the sixth ligand with the change in the transition energy of the visible band shown in Table 3.1. (Assume that the $Cr(CO)_5$ z^2 level is located at -6.75eV.)

Table 3.1 Dependence of the visible band wavelength on the nature of X in $Cr(CO)_5X$.

X	absorption wavelength (nm)
Ne	624
SF_6	560
CF_4	547
Ar	533
Kr	518
Xe	492
CH_4	489

*3.15. $MX_3(NMe_3)_2$ species (X = halide, M = Ti, V, Cr) have distorted geometries (C_{2v}) whereas a trigonal bipyramidal (D_{3h}) structure might have been expected. The magnitude of the distortion decreases in the order $Cr > Ti > V$. Why is this?

3.16. Use the angular overlap model or otherwise to construct a molecular orbital diagram for a square-planar ML_4 complex. Show how the form of the diagram depends upon whether the ligands are π-acceptors or donors. The electronic spectra of several complexes of this type may be well-fitted using the two parameters e_σ and e_π. However the energy of the z^2 orbital is usually not fitted well by these parameters. The transition from z^2 to $x^2 - y^2$ occurs at a much higher energy than expected. By considering the symmetry properties of the $(n+1)s$ and p orbitals provide an explanation for this observation.

3.17. (a) Draw a Walsh diagram for the square planar to tetrahedral interconversion in ML_4 for only the five metal-based d orbitals. There is no need to draw the orbitals, just label them according to the Mulliken scheme and the type of d orbital (x^2-y^2, z^2, etc.) using the coordinate system shown in 3.6. (b) Predict the geometry for each of the molecules $Ni(CO)_4$, $PtCl_4^{2-}$,

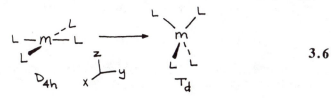

3.6

$Fe(CH_3)_4$.

*‡3.18. $MoO_2(CN)_4^{4-}$ ions are known to be octahedral and have been structurally characterized in the salt $NaK_3[MoO_2(CN)_4].6H_2O$. (a) Construct molecular orbital diagrams for both the *cis* and *trans* isomers of this ion, paying particular attention to the influence of π bonding of the splitting of the *xy*, *xz*, and *yz* orbitals. Be sure to indicate the influence of both the π acceptors and π donors. (b) Assuming for simplicity that the influence of the π donors is dominant, predict which isomer will be most stable for this ion. (c) Predict the structure of the hypothetical $MoO_2(CN)_4^{2-}$ ion. (You may have to algebraically solve the relevant 2x2 determinant(s) to get your answer.) (d) Repeat the calculation using the Angular Overlap Model. (e) What is the general rule that comes out from your orbital pictures? (f) Use similar arguments to comment on the electron count dependence of the relative stability of the possible conformers of *trans* bisethylene and *trans* biscarbene compounds of the transition metals.

3.19. The major contributions to the ^{59}Co chemical shifts of low spin Co^{III} compounds are thought to be (a) a diamagnetic term largely independent of the ligand, and (b) a term resulting in temperature independent paramagnetism (*TIP*) *via* a mixing of ground and excited electronic states. The latter from perturbation theory should increase inversely with the energy difference between the ground and first excited electronic states. In this case the latter is the lowest energy *d-d* transition. Decide how the chemical shift should vary with the nature of the ligands, CN^-, NH_3 and H_2O. The chemical shift for the molecule containing the tripod ligand $L^- = [(C_5H_5)Co\{P(O)(OR)_2\}_3]^-$ ($R = C_2H_5$ etc.,) is larger than any other. What might be a rather special property of this species?

3.20. By using p orbitals on the bridging oxygen atoms only generate the energy diagram for the interaction of the copper '*xy*' orbitals with the bridging p orbitals in the complex shown in **3.7**. Show how the relative energy of these two orbitals is expected to change with the angle α. Whether one finds the

3.7

singlet or triplet state associated with the two 'copper' electrons is determined by a balance of the one- and two-electron terms in the energy.

$$\Delta E(S,T) = 2K_{ab} - \Delta^2/U$$

K is the exchange integral and is always positive leading to a ferromagnetically coupled system (triplet state lowest in energy), but the second term favors an antiferromagnetic arrangement (singlet state lowest in energy). Δ is approximately the energy difference between the two orbitals you have derived and $U = J_{aa} - J_{ab}$ which measures the difference in Coulombic repulsion for the cases of two electrons located on the same center and of two electrons located on adjacent centers. Assuming that K_{ab} and U remain invariant to changes in α, rationalize the observation that a crossover from an antiferromagnetic to a ferromagnetic state occurs when α decreases from 105° to 96°.

3.21. What would be the geometry of a d^8 PtL_4 species in its first excited electronic state? Draw an orbital correlation diagram to rationalize the photochemical *cis-trans* isomerisation observed in several $PtL_2L'_2$ compounds.

3.22. One possible geometry for the MH_8 molecule, where M is a transition metal, is the D_{6h} structure of **3.8**. For simplicity, assume that all M-H bond

3.8

lengths are equal at 1.41Å. The H_1-H_2 distance is then 1.99Å but the H_2-H_3 distance short at 1.41Å. (In free H_2 it is 0.74Å.) Hydrogen will be a little more electronegative than the transition metal. Draw out an orbital interaction diagram for this molecule. On one side, draw out the Symmetry Adapted Linear Combinations of the H_8 group and on the other side, draw out the nd, $(n+1)s$ and $(n+1)p$ atomic orbitals of the metal. Label each resultant molecular orbital and draw out its shape. Finally, predict which electron counts should yield a stable molecule (there will be two) and propose what element M should be (do not consider lanthanide and actinide possibilities).

‡*3.23. The metal-ligand bond strength in a complex depends upon the other ligands present and is often very strongly influenced by the nature of the ligand *trans* to it (the *trans*-influence). For example, the Cr-CO distance in $Cr(CO)_5L$ is 1.95Å if $L = CO$ but shorter (1.86-1.88Å) if L is PPh_3, $P(OPh)_3$ or $P(CH_2CH_2CN)_3$. This is sometimes viewed rather vaguely in terms of competition for the available electron density at the metal by the two ligands. Similar result are found when the ligands are π-donors. A simple molecular orbital result will put this on a more quantitative footing. Set up a secular determinant which represents the interaction of the two *trans* π orbitals with a central metal d orbital of π type. Choose the situation where the ligands are π-donors.The problem is simplified if a symmetry adapted ligand combination is used. Write down the quadratic equation which describes the orbital energies in terms of H_{ML}, the metal-ligand interaction integral, and H_{MM} and H_{LL}, the metal and ligand orbital ionization energies. Solve the quadratic by expansion as a power series in $H_{ML}/(H_{LL}-H_{MM})$ and show by consideration of the fourth order terms that the stabilization energy per ligand is smaller than that for the case where only one such ligand is coordinated. (Alternatively you may use a computer program to solve the secular determinant, either numerically or

symbolically.)

3.24. For what transition metal ML_n geometry is a three above two d orbital splitting pattern found with symmetry labels t_{2g} and e_g respectively? How is Δ related to Δ_{oct} for this geometry?

3.25. It has been suggested that the compound shown in **3.9** (where L is a two-electron ligand) actually exists in two forms. The X-ray structure of the two forms suggests that there is little difference in the basic structure, the major change is associated with the $W\text{-}O$ bond distance. In one molecule it is found

3.9

to be 1.72Å, whereas in the other it is 1.89Å. (a) By means of a Walsh diagram show why these two different molecules could in principle co-exist and why there is a sizable barrier that interconverts them. (Hint, O is a stronger π donor than Cl). (b) Describe what should happen to the two $W\text{-}Cl$ bond lengths on going from one to the other. (In some of these compounds L is a phosphine, in others $L3$ is the tridentate amine ligand; $Me_2N\text{-}CH_2CH_2\text{-}N(Me)\text{-}CH_2CH_2NMe_2$.)

*3.26. (a) Using a d-orbital only model the two short-four long and two long-four short distortions of a Jahn-Teller unstable octahedron (*e.g.*, d^9) turn out to be very similar in energy. Use the angular overlap model, assume that the change in e_σ is proportional to displacement, for one component of the Jahn-Teller active mode of e_g symmetry to show that the distortion energies are in fact identical for such a simple scheme. (b) Almost all Jahn-Teller unstable Cu^{II} environments distort to give the two long-four short geometry. In a qualitative way include the unoccupied $(n+1)s$ orbital into your model, and show how d-s mixing on distortion accounts for the observed distortion. (c) Now extend your model to show how the two short-four long distortion, so characteristic of Hg^{II} chemistry, arises. For this element linear two-coordination is almost universally found for ligands with an electronegativity lower than that of sulfur.

3.10   **3.11**   **3.12**

*3.27. The structure of **3.10** is square planar, *i.e.*, the N-Os-N and P-Os-P angles are 180°. Construct a generalized Walsh diagram for the square planar to tetrahedral transformation in X_2ML_2 molecules (where X = NR or O) to explain this observation. In both compounds R and R' are alkyl or aryl groups. In compound **3.11** the O-Re-O angle is around 135° and the R-Re-R angle about 85° for a whole series of molecules of this type. How does this fit into your scheme? The molecule **3.12** is also square planar. Predict the conformation of the OR groups for this to be so in your model.

3.28. By using molecular orbital arguments rationalize the observation that shorter metal-ligand distances are found in low-spin square planar Ni^{II} and Co^{II} molecules than in comparable high-spin tetrahedral analogs.

3.29. The M-O bond lengths for a series of $M(H_2O)_6^{2+}$ complexes are: $V(H_2O)_6^{2+}$, V-O = 2.14Å; $Mn(H_2O)_6^{2+}$, Mn-O = 2.16Å; $Fe(H_2O)_6^{2+}$, Fe-O = 2.13Å; $Ni(H_2O)_6^{2+}$, Ni-O = 2.04Å. All molecules are high spin. Provide a molecular orbital rationale for this order. A typical pattern across a series of main groups is CH_4, C-H = 1.09Å; NH_3, N-H = 1.01Å; OH_2, O-H = 0.96Å, H-F = 0.92Å).

3.30. Adapt Figure 17.11 from Reference 1 so that it shows a Walsh diagram for Fe-O-O angle bending in a heme unit for oxyhemoglobin (**3.13**). For this purpose, just consider σ_g and π_g in O_2 as analogs of n and π^* in NO. Also, put the iron $x^2 - y^2$ orbital in the diagram just below π^*- λyz but above $z^2 - \lambda n$. For the ultimate low spin "product" count O_2 as O_2^{2+}, thus the electronic configuration of O_2 is . . . $(\sigma_g)^2(\pi_u)^4$. Determine whether the Fe-O-O angle is linear or bent. Consider the initially formed high spin complex when O_2 in the . . . $(\sigma_g)^2(\pi_u)^4(\pi_g)^2$ configuration interacts with a high spin heme unit in deoxyhemoglobin. Predict the Fe-O-O angle.

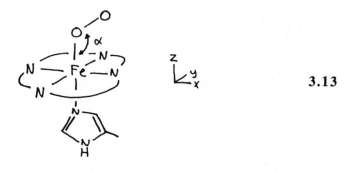

3.13

Answers

3.1. We know that for a complex containing n_λ ligand orbitals of a particular type, λ ($\lambda = \sigma, \pi$) then the sum total interaction energy with all the d orbitals is equal to $n_\lambda e_\lambda$ using the quadratic parameter e_λ. (Note a similar rule does not hold for the quartic term f_λ.) In the octahedron the σ interaction is associated with the e_g pair of d orbitals. Thus the quadratic destabilization (3.14) of each pair is $n_\sigma e_\sigma/2 = 6e_\sigma/2 = 3e_\sigma$. In the tetrahedron the σ interaction is associated

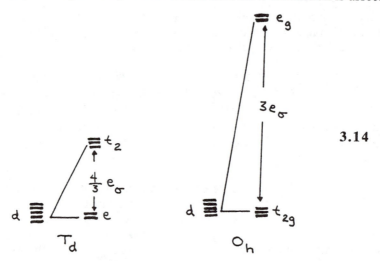

3.14

with the t_2 orbitals. In this case the destabilization energy is $n_\sigma e_\sigma/3 = 4e_\sigma/3$. Thus using the σ-only model, and only the quadratic term in the *AOM*, $\Delta_{oct} = 3e_\sigma$ and $\Delta_{tet} = 4/3e_\sigma$ so that $\Delta_{tet} = 4/9\Delta_{oct}$. When the quartic *AOM* terms or the effects of π bonding are introduced this simple relationship does not hold. A further assumption of the model is that e_σ is equal for both geometries. This is usually not true since bond lengths and thus overlap integrals vary with metal-ligand distance.

Given two electrons and two orbitals the factors (See Chapter 8 of Reference 1) which determine the spin state of the system are the orbital energy separation (e_1-e_2) which favors the singlet and the Coulombic repulsion between the two electrons in the same orbital (J_{aa}) which favors the triplet. If the orbital energy separation is small then perhaps (e_1-e_2) $< J_{aa}$ and the triplet high spin situation will be stable. For large energy separations then the singlet (low spin situation) will be favored. Since (e_1-e_2) ~ $\Delta_{tet} = 4/3e_\sigma$, low spin tetrahedral complexes are expected to be less prevalent than low spin octahedral ones where (e_1-e_2) = $3e_\sigma$.

The *MOSE* values for high and low spin d^8 configurations are easily calculated and are shown in Table 3.2.
Notice that the *MOSE* values of several of these systems are equal at $6e_\sigma$. This suggests that there are not enormous energetic barriers between these geometries for these specific electron counts and that they may be readily

Table 3.2. *MOSE* values for high and low spin d^8 configurations

	octahedral	square pyramidal	square
high spin	$6e_\sigma$	$5e_\sigma$	$4e_\sigma$
low spin	$6e_\sigma$	$6e_\sigma$	$6e_\sigma$

interconverted.

3.2   **3.15** shows the construction of the molecular orbital diagram for the square pyramid using σ orbitals on the ligands. The effect of interactions of π type will depend on the donor or acceptor nature of the ligands. Donors will

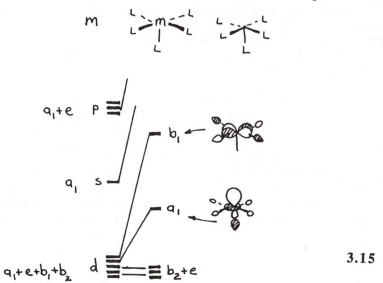

3.15

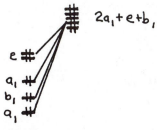

push the $b_2 + e$ set up in energy, acceptors the converse. With five strong field CN^- ligands a large $b_2 + e / a_1 + b_1$ gap is expected and a low-spin electronic ground state is expected, but with oxygen or nitrogen donors such as the ligand in the question, this splitting will be much smaller and a high-spin arrangement will be preferred. The electronic configurations for the two are thus $...(z^2)^2$ and $...(z^2)^1(x^2-y^2)^1$ respectively. The geometrical differences stem from this. Notice that from **3.15** x^2-y^2 is only involved in σ antibonding interactions with the basal ligands whereas z^2 is involved with both apical and basal ligands. With equal occupation of these two σ antibonding orbitals the apical and basal bonds should be of approximately equal strength (and lengths) as they are in the high spin complex. With double occupation of z^2 there is little weakening of the basal linkages (antibonding interaction with the collar) but significant weakening of the apical linkage as a result of strong antibonding interaction with the lobe of z^2. The metal-apical distance in $Ni(CN)_5^{3-}$ is very long indeed.

These results prompt speculation concerning the bond lengths in molecules where all the σ antibonding orbitals are occupied as in d^{10} Zn^{II} systems for example. Are all the bond lengths long here? The answer is no. On moving from left to right across the periodic table the bond lengths in general get shorter (Question 3.29) and by the time Zn^{II} is reached the occupied nd orbitals are largely core in character. The $(n+1)s$, and p orbitals are now responsible for chemical bonding.

3.3   See Question 7.8. For the electronic configurations appropriate for both systems the symmetrical structure lies at a local energy maximum.

3.4   The generation of this molecular orbital diagram is such a frequently asked question that we will not describe the answer here but refer the reader to almost any inorganic text. Two diagrams which show the important differences between π-donors and acceptors are shown in **3.16**. Notice that in the donor case the t_{2g} orbitals are destabilized and are M-L antibonding whereas in the acceptor case they are stabilized and M-L bonding. (a) Thus $Fe(CN)_6^{4-}$ has a

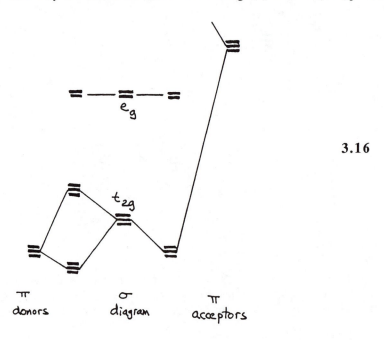

3.16

shorter *Fe-C* distance than $Fe(CN)_6^{3-}$ because the extra electron resides in a *M-C* bonding orbital, the CN^- ligand being a π-acceptor. (b) However on going from d^3 for *M* = *Cr* to d^5 for *M* = *Fe*, the extra electrons in these high-spin species occupy the *M-L* σ antibonding e_g levels and the *M-Cl* stretching frequency decreases on going from *M* = *Cr* to *M* = *Fe*. The case of *M* = *Mn* is a tricky one because, with one electron the e_g orbital, the molecule is Jahn-Teller unstable, and more than one *IR*-active *M-Cl* stretch is seen.

3.5   (a) The diagram is constructed in **3.15**. (b) The movement of the d orbitals is shown in **3.17**. As a result of the reduction of overlap on bending

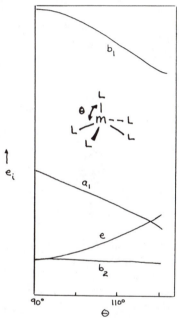

3.17

the z^2 and $x^2 - y^2$ orbitals drop in energy since they become less antibonding. However, the *xz*, *yz* pair, nonbonding at α = 90°, become antibonding, and thus increase in energy as α increases. (c) From this diagram we can see that the low-spin complex will prefer a geometry close to 90°, but the high-spin analog a geometry with a higher value of α as a result of the stabilization of the now occupied z^2 and $x^2 - y^2$ orbitals. This is a more staisfying explanation of the geometrical results than one in the literature, which suggests that high-spin iron is too large to lie in the plane of the four nitrogen atoms of the heme unit but size factors are fine for the low-spin form. (d) For this part of the question we need a correlation diagram for the square pyramid (*spy*) to trigonal bipyramid (*tbp*), $C_{4v} \leftrightarrow D_{3h}$ interconversion This is shown in **3.18**. Notice the change in the direction of the *z* axis between the two. The angular overlap energies are given for the σ orbitals. It is immediately apparent that a singlet d^6 $Cr(CO)_5$ molecule is Jahn-Teller unstable at the *tbp* geometry with the electronic configuration $(e'')^4(e')^2$ but not at the *spy* structure. The singlet states for a d^6 $Cr(CO)_5$ molecule are found by evaluation of the symmetric direct product of e' to be $^1A_1' + {}^1E'$. The $^1E'$ state is Jahn-Teller unstable, the

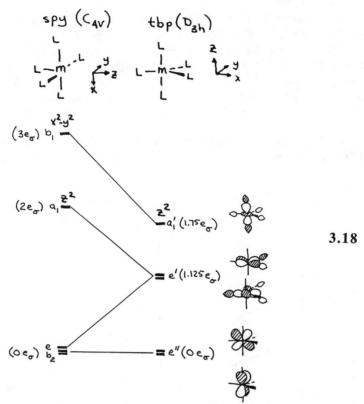

symmetry species of the Jahn-Teller active mode being given again by the symmetric direct product of e' as $a_1' + e'$. One component of the e' bending vibration, shown in **3.19**, takes the *tbp* to the *spy*. The d^8 $Fe(CO)_5$ molecule is Jahn-Teller stable at both geometries. In terms of the angular overlap model the energy difference between *spy* and *tbp* is $0.5e_\sigma$ for $Fe(CO)_5$ (with the *spy* predicted to be more stable) but $2.5e_\sigma$ for $Cr(CO)_5$. Thus we might expect that

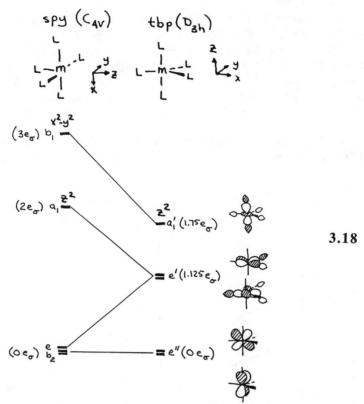

the two geometries have similar energies for $Fe(CO)_5$. A more sophisticated calculation is needed to reproduce the rather small (~1kcal/mole) energy difference between the two structures.

3.6.   Consider the molecular orbital diagram for the substituted octahedral molecule shown in **3.20**. Shown is the case for a *trans* MX_4YZ molecule where the σ strength of Y and Z is less than that of X. The result is that z has dropped below the $x^2 - y^2$ orbital. We have chosen Z to be the better σ donor

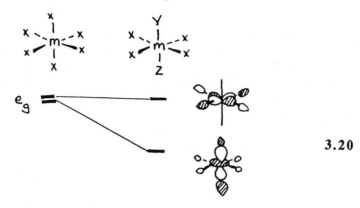

3.20

and thus it has a larger orbital coefficient in z^2. Photochemical excitation leads to population of the lowest excited state orbital- in this case z^2, the orbital which is dominated by the 'weaker field' ligands, Y and Z. Examination of its properties gives us information concerning the relative bond strengths in the excited state. Clearly is is only weaky antibonding between M and X, but is most strongly antibonding between M and Z. This then, the 'stronger field ligand' is the ligand which will be ejected.

3.7. First we point out that the trigonal bipyramidal geometry is Jahn-Teller unstable for $Cr(CO)_5$ (see Question 3.5) as a singlet. (A triplet state with this geometry is expected to be stable.) The Jahn-Teller active mode of the trigonal bipyramid for the d^6 configuration is of species e' and as may be seen from **3.19**, both opening and closing motions are in principle favored routes. The angular overlap model gives energies of $10e_\sigma$ (spy) $7.75e_\sigma$ (tbp) and $\sim 8.7e_\sigma$ (C_{2v} structure). To get the energy of the latter a secular determinant needs to

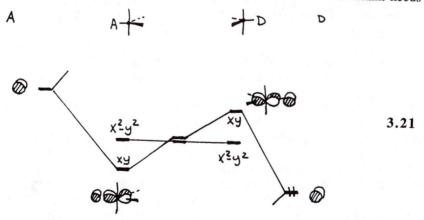

3.21

be solved. (See reference 2 page 151). Thus on σ grounds the *tbp* structure should distort to the *spy* (**3.18**). Now the *e'* degeneracy of the *tbp* structure (see **3.18** for a molecular orbital diagram) may be split by introduction of a π ligand *trans* to the angle α. Clearly (**3.21**) π donors destabilize *xy* such that the crossing occurs at higher α (**3.22**) and π acceptors stabilize *xy* such that the crossing occurs at lower α (**3.23**). If this π effect is sufficient to overwhelm

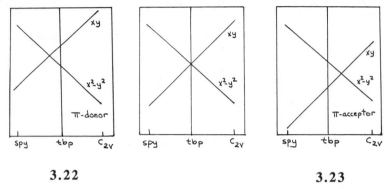

3.22   **3.23**

the σ effect described using the *AOM* then π donors (*e.g.*, *Cl*) will tend to give smaller angles α (*C₂ᵥ*), whereas π-acceptors larger angles (*spy*). The C_{2v} structure is probably a transition state for the thermal rearrangement of *Cr(CO)₅*.

3.8   The *CO* stretching frequency (or rather the *CO* force constant) is a measure of the extent of mixing between $CO\pi^*$ and the filled metal *d* orbitals (see Question 5.3). The larger the mixing of $CO\pi^*$ into the occupied metal levels, the lower the *CO* bond order and the lower the *CO* stretching frequency. From simple perturbation theory arguments the smaller the energy separation between the metal d levels and the π^* orbitals shown in **5.23** the larger the interaction. Now an increase in the positive charge on the metal leads to an increase in its ionization potential and an increase in its negative charge the converse. Thus ΔE varies for the three (isoelectronic) molecules under consideration; it is largest for the most highly positively charged case $(Mn(CO)_6^+)$ and smallest for the most negatively charged molecule $(V(CO)_6^-)$. Thus the stretching frequencies change in the order given.

3.9   The relevant orbital correlation diagram is shown in **3.18** in the answer to Question 3.5. and is extended for the present problem in **3.24**. The location of the electrons are shown at various stages of the Berry process which involves interconversion of the different square pyramidal (*spy*) *Cr(CO)₅* units *via* a trigonal bipyramid (*tbp*). Notice that on excitation of a *spy* *Cr(CO)₅* molecule, one electron is promoted from the *e* set to the z^2 orbital. The *spy* geometry is unstable for this electronic configuration, and distorts to a trigonal bipyramid. Such a singlet *tbp* *Cr(CO)₅* is unstable at this geometry too but can smoothly relax back to a ground state *spy* molecule. It may (**3.25**) either relax to the starting *spy* structure (with the ligand indicated by • in a basal position) or one of two other possibilities (one of which has the ligand • in the apical

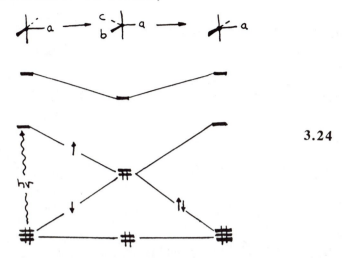

3.24

position). This provides a route for apical/basal ligand exchange. The apical position of each of the square pyramids is labeled with an *a*.

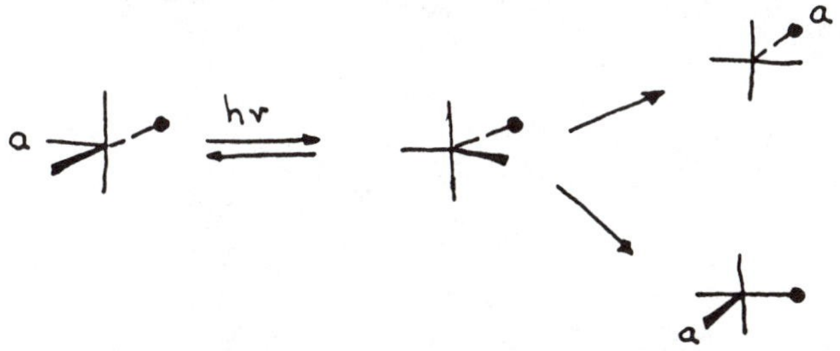

3.25

3.10. We will separate the hydrogen atom orbitals into two sets, those associated with the trigonal prism (set 1, χ_1-χ_6) and those with the capping atoms (set 2, χ_7-χ_9) as shown in **3.26** and **3.27** respectively. The basis

3.26 H_1 H_3 H_2 + H_9 H_7 **3.27**
                    H_4 H_6 H_5 H_8

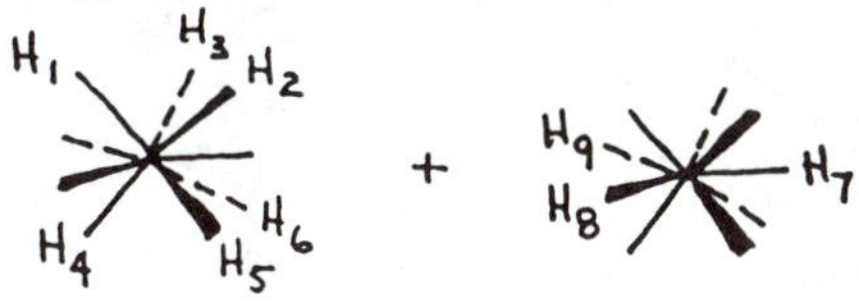

orbitals χ_1-χ_6 transform as $a_1' + e' + a_2'' + e''$ under the D_{3h} point group, and the orbitals χ_7-χ_9 as $a_1' + e'$, as shown in Table 3.3

Table 3.3. Transformation Properties of the Basis Orbitals

D_{3h}	E	$2C_3$	$3C_2$	σ_h	$2S_3$	$3\sigma_v$	
Γ_1	6	0	0	0	0	2	$a_1' + e' + a_2'' + e''$
Γ_2	3	0	1	3	0	1	$a_1' + e'$

$$\psi_{a'} \propto \chi_1 + \chi_2 + \chi_3 + \chi_4 + \chi_5 + \chi_6 \equiv$$

$$\psi_{e'}^{(1)} \propto 2\chi_1 - \chi_2 - \chi_3 + 2\chi_4 - \chi_5 - \chi_6 \equiv$$

$$\psi_{e'}^{(2)} \propto \chi_2 - \chi_3 + \chi_5 - \chi_6 \equiv$$

$$\psi_{a_2''} = \frac{1}{\sqrt{6 + 12S_1}} (\chi_1 + \chi_2 + \chi_3 + \chi_4 + \chi_5 + \chi_6) \equiv$$

3.28

$$\psi_{e''}^{(1)} = \frac{1}{\sqrt{12 - 12S_1}} (2\chi_1 - \chi_2 - \chi_3 - 2\chi_4 + \chi_5 + \chi_6) \equiv$$

$$\psi_{e''}^{(2)} = \frac{1}{\sqrt{4 - 4S_1}} (\chi_2 - \chi_3 - \chi_5 + \chi_6) \equiv$$

$$\psi_{a'} \propto \chi_7 + \chi_8 + \chi_9 \equiv$$

$$\psi_{e'}^{(1)} \propto 2\chi_7 - \chi_8 - \chi_9 \equiv$$

$$\psi_{e'}^{(2)} \propto \chi_8 - \chi_9 \equiv$$

The orbitals of each type are drawn out in **3.28**. Here the overlap integrals are $S_1 = \langle \chi_1 | \chi_2 \rangle$, $S_2 = \langle \chi_1 | \chi_8 \rangle$ and $S_3 = \langle \chi_6 | \chi_7 \rangle$. We may usefully construct new combinations of the a_1' and e' orbitals associated with the two sets as shown in **3.29**. These sets, although normalized are not exactly orthogonal since the S_i are not all equal. Notice that when we take linear combinations of the e' sets the orbital picture looks a little complex. There is more antibonding character in $\psi_{1e'}^{(1)}$ than in $\psi_{2e'}^{(1)}$. The orbitals χ_1 and χ_8, χ_1 and χ_9, χ_4 and χ_8, etc., are antibonding in $\psi_{1e'}^{(1)}$ but bonding in $\psi_{2e'}^{(1)}$. In $\psi_{2e'}^{(2)}$ however, there is bonding between χ_2 and χ_8, χ_3 and χ_9, χ_5 and χ_8, χ_6 and χ_9. The energetic ordering of these H_9 orbitals follows directly from the number of nodal planes and the extent of overlap between pairs, as shown in **3.30**. Now we may construct the molecular orbital diagram for the MH_9 species as shown in the same diagram.

$$\psi_{1a_1'} = \frac{1}{\sqrt{9+12S_1+24S_2+6S_3}}\left(\chi_1+\chi_2+\chi_3+\chi_4+\chi_5+\chi_6+\chi_7+\chi_8+\chi_9\right)$$

$$\equiv$$

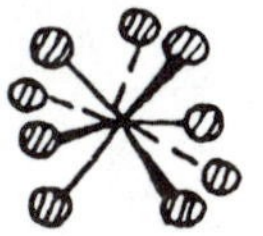

$$\psi_{2a_1'} = \frac{1}{\sqrt{9+12S_1-24S_2+6S_3}}\left(\chi_1+\chi_2+\chi_3+\chi_4+\chi_5+\chi_6-\chi_7-\chi_8-\chi_9\right)$$

$$\equiv$$

$$\psi_{1e'}^{(1)} = \frac{1}{\sqrt{18-12S_1-24S_2-6S_3}}\left(2\chi_1-\chi_2-\chi_3+2\chi_4-\chi_5-\chi_6+2\chi_7-\chi_8-\chi_9\right)$$

$$\equiv$$

3.29

$$\psi_{1e'}^{(2)} = \frac{1}{\sqrt{6-4S_1-8S_2-2S_3}}\left(\chi_2-\chi_3+\chi_5-\chi_6-\chi_8+\chi_9\right)$$

$$\equiv$$

$$\psi_{2e'}^{(1)} = \frac{1}{\sqrt{18-12S_1+24S_2-6S_3}}\left(2\chi_1-\chi_2-\chi_3+2\chi_4-\chi_5-\chi_6-2\chi_7+\chi_8+\chi_9\right)$$

$$\equiv$$

$$\psi_{2e'}^{(2)} = \frac{1}{\sqrt{6-4S_1+8S_2-2S_3}}\left(\chi_2-\chi_3+\chi_5-\chi_6+\chi_8-\chi_9\right)$$

$$\equiv$$

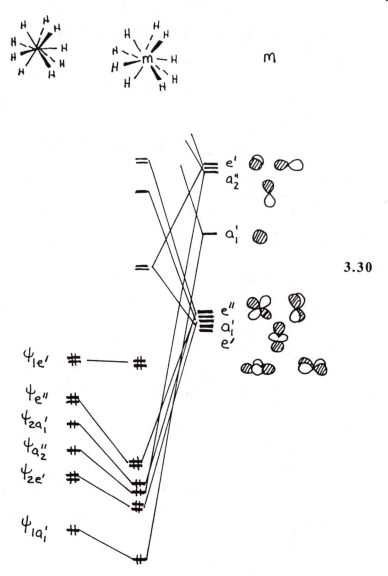

3.30

3.11. Loss of two *trans* ligands from the octahedron leads to a drop in the energy of the z^2 orbital resulting from the loss of two strong antibonding interactions (**3.31**). If the four equatorial bonds shrink as a result then the $x^2 - y^2$ orbital will be destabilized. To obtain the electronic states of the d^8 configuration (e_g^2) we need to evaluate the symmetric and antisymmetric direct products of e_g. This is done for the point group O (this is good enough for our needs since we know $g \times g = g$) in Table 3.4.

Table 3.4. The symmetric and antisymmetric direct products of e in the point group O.

for e	E	$6C_4$	$3C_2(=C_4{}^2)$	$8C_3$	$3C_2$	
$\chi(\mathcal{R})$	2	0	2	-1	0	
$\chi(\mathcal{R}^2)$	2	2	2	-1	2	
$\chi^2(\mathcal{R})$	4	0	4	1	0	
$\chi^2{}_{sym}(\mathcal{R})$	3	1	3	0	1	$a_1 + e$
$\chi^2{}_{antisym}(\mathcal{R})$	1	-1	1	1	-1	a_2

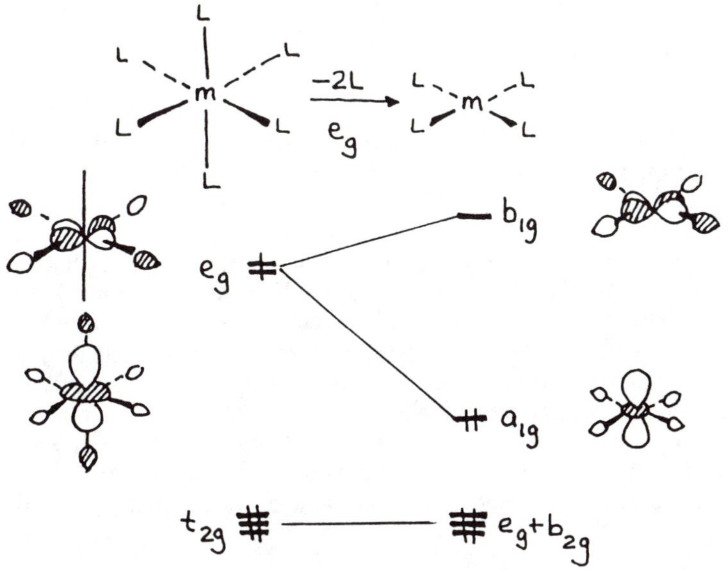

3.31

This leads to the electronic states $^3A_{2g}$, $^1A_{1g}$ and ^1E_g. Recalling Hund's rules for atoms (although we should be careful when applying them to molecules) the triplet will probably lie lowest in energy and the doubly degenerate singlet next in energy. This ^1E_g state is Jahn-Teller unstable and splits apart in energy during an e_g symmetry vibration, *i.e.*, the same motion shown in **3.31**. If the $x^2 - y^2$ orbital is destabilized more than the z^2 orbital is stabilized on distortion then the energetic behavior of the relevant electronic states is shown in **3.32**. The energy difference between the two electronic states at the left hand side of the diagram is set by the two-electron terms in the energy. These are the Coulomb and Exchange terms which depend upon $1/r_{12}$, the interelectronic separation. The energy changes on moving towards the right hand side of the diagrams is controlled by the changes in overlap and the energies of the orbitals shown in **3.31**. Two situations are shown. At the left (a) in **3.32** the singlet-triplet difference is small, the case for heavier transition metals. Here the singlet crosses the triplet somewhere along the distortion coordinate. Thus we expect for this case (found for Pt^{II} in $Pt(CN)_4{}^{-2}$) to find a low-spin square planar molecule. On the right side of **3.32**, (b), a possibility for a first row

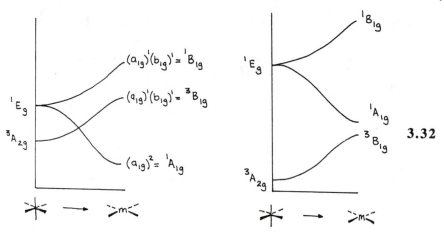

3.32

transition metal, we expect to find the octahedral triplet, as in $Ni(H_2O)_6^{+2}$. Behavior of the first type is also found for first row transition metals too. (Also see question 8.3.)

3.12. The obvious starting geometry for this species is the trigonal plane. A molecular orbital diagram for this geometry is readily obtained from that of the ML_5 trigonal bipyramid by the removal of two *trans* ligands as in **3.33**. What is not easy to see, except by calculation, is that the z^2 orbital has dropped below the $x^2 - y^2/xy$ pair in the process. We show the angular overlap energies for the relevant orbitals which can be used to substantiate this. The result of this orbital crossing is that a low spin d^8 species is Jahn-Teller unstable in the D_{3h} geometry. The distortion mode which relieves this degeneracy is obtained from the symmetric direct product of e'. This is also of symmetry e', and a

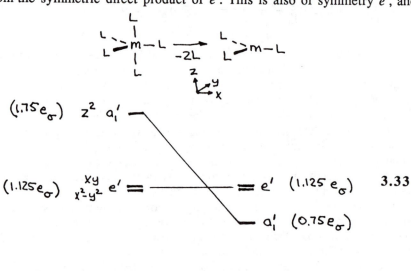

3.33

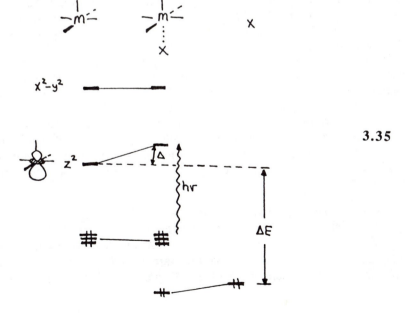

3.34

motion of this type takes the trigonal plane to the T and Y shapes (**3.34**). The structure is in fact intermediate between the D_{3h} geometry and the T shape with an angle α of around 154°. It is important to note that distortion to a pyramidal structure from the D_{3h} geometry does not relieve the Jahn-Teller instability.

3.13. From the molecular orbital diagram of **3.18** we can see that for the low-spin d^8 configuration the only occupied σ antibonding orbitals are those which are antibonding between the metal and the equatorial ligands. The prediction is therefore that the axial bonds will be stronger than the equatorial ones. In fact the situation is a little more complex than this and the interested reader is referred to *Inorg. Chem.*, **14**, 365 (1974).

3.14. (a) The construction of the molecular orbital diagram is shown in **3.35** and stresses the interaction of an orbital located on the sixth ligand with the empty z^2 orbital of the square pyramidal fragment. The lowest energy d-d transition is from the e pair to this a_1 orbital and its energy will depend upon the size of the shift Δ which arises from the interaction of z^2 with the sixth site occupant. (b) Perturbation theory suggests that the interaction energy will depend inversely upon the energy separation ΔE of **3.35**. This is readily

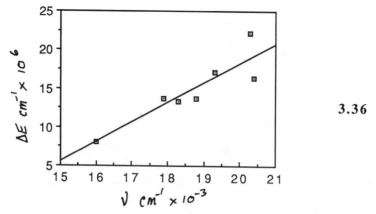

3.36

evaluated in the following way. The energy of the X located orbital can be approximated as being close to the ionization potential of the species concerned. The energy of the empty z^2 orbital is more difficult to estimate but we chose 6.75eV. (In fact the correlation is not very sensitive to this value.) **3.36** shows the correlation between the transition energy and this parameter $1/\Delta E$. The agreement is quite reasonable.

3.15. **3.18** shows the molecular orbital diagram for a trigonal bipyramidal MX_5 molecule, which is also appropriate for the MX_3Y_2 molecule of D_{3h} symmetry. The e' levels are both π and σ antibonding, the e'' levels just π antibonding. The electronic configurations for the Ti, V and Cr systems are thus $(e'')^1$, $(e'')^2$ and $(e'')^2(e')^1$ respectively. High-spin Cr^{III} is expected here since weak-field ligands are coordinated. Notice that both the Ti and Cr systems are Jahn-Teller unstable with $^2E''$ and $^4E'$ states respectively. The Jahn-Teller active coordinate is obtained by evaluating the symmetric direct product of e' and e''. In both cases it is $a_1' + e'$, leading to e' as the symmetry species of the distortion coordinate. This will take the D_{3h} to a C_{2v} structure. However the instability for the Ti case is associated with a degeneracy in orbitals associated with π-type interactions and is expected to be be smaller than that for the Cr case where the degeneracy is associated with orbitals involved in σ interactions. Thus the magnitude of the distortion should be in the order $Cr > Ti$, as observed. No orbital degeneracy is found for the d^2 vanadium case ($^3A_2'$) and yet a small distortion is found in the solid. Such small distortions are usually explained as arising from the ubiquitous 'crystal packing forces'. These are used to 'explain' solid-state distortions which cannot be understood using electronic arguments. Such small distortions are always possible if, as in the present case, the metal atom does not sit in a site of high enough symmetry.

3.16. The molecular orbital diagram is assembled in **3.37**. It is a simple matter to derive the relevant interaction energies using the AOM. The $x^2 - y^2$ orbital is destabilized by $3e_\sigma$ and z^2 by e_σ. The π interactions are $4e_\pi$ (b_{2g}) and $2e_\pi$ (e_g). Different ordering patterns are found for π donors and acceptors. One set of interactions left off the diagram are those with the $(n+1)s$

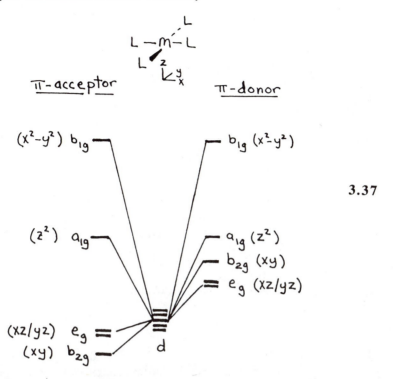

and $(n+1)p$ orbitals on the metal. We may add these in now. Since in this point group the d and p orbitals are of opposite parity with respect to inversion they cannot mix, but the $(n+1)s$ orbital is of the same symmetry (a_{1g}) as z^2. Interaction between the two depresses z^2, often by an amount close to the size of e_σ itself, leaving the xy orbital as the *HOMO* of the d^8 molecule in the π donor case **(3.38)**.

3.17. (a) The correlation diagram is shown in **3.39**. (b) For $Ni(CO)_4$ (d^{10}) all metal d-ligand bonding and antibonding orbitals are filled, and it is not immediately clear from the diagram which geometry should be adopted. However, if we ignore the d orbitals for the moment by considering them to be core orbitals, then the molecule is just like CH_4, except that here it is the $4s$ and $4p$ orbitals that are used on nickel. Using this analogy we would expect a tetrahedral geometry. There is an electronic observation which gives the same result. The most stable stucture is found when each of the filled molecular orbitals describe interactions which are as equal as possible between central atom and ligand. Clearly in the tetrahedral geometry the interactions with the central atom p orbitals are equal by symmetry. A similar rule is found in many other places in chemistry, including that of H_2O_2 of Question 4.32, CH_4 itself of Question 4.14 and the *cis* and *trans* dioxocomplexes of Question 3.18. Alternatively one can easily show from a Walsh diagram that of the four M-L σ-bonding orbitals, one is greatly destabilized on going from T_d to D_{4h}. $PtCl_4^{2-}$ is a low-spin d^8 system and the square planar geometry is the one where all of the electrons lie in stable molecular orbitals. $Fe(CH_3)_4$ would have

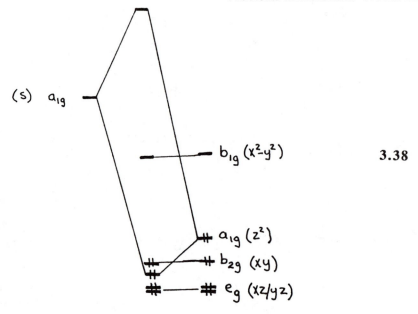

3.38

a d^4 configuration. If low-spin, a closed shell arrangement would exist for the tetrahedral geometry. If high-spin, the obvious arrangement is the square planar one.

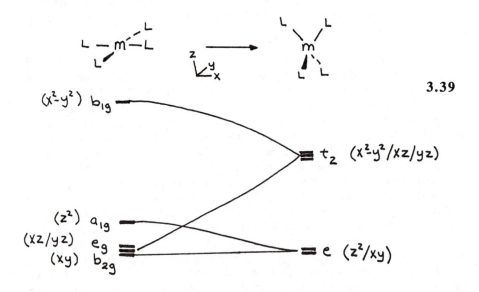

3.39

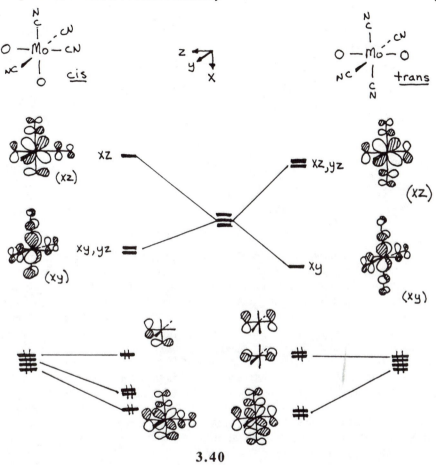

3.40

3.18. (a) The diagrams are constructed in **3.40**. For the *trans* (higher symmetry) isomer the *xz* and *yz* orbitals are involved in two π donor and two π acceptor interactions. (Only one is shown pictorially.) If the donor interaction is larger than the acceptor interaction then this pair is destabilized relative to the free *d* orbitals. The *xy* orbital is stabilized as a result of four stabilizing π acceptor interactions. A similar pattern is found for the largely oxygen located π orbitals. Since only two of the *d* orbitals contain oxygen character, only two oxygen combinations are stabilized as shown. There are a pair of orbitals which cannot interact by symmetry and remain as lone pairs. The splitting pattern is the converse in the *cis* isomer. Here the *xy*, *yz* pair are stabilized (three acceptor interactions and one donor) and the *xz* orbital is destabilized (two donor and two acceptor interactions). Here there are three orbitals which can interact with the oxygen atoms, leading to three oxygen located bonding orbitals. (b) Assuming a weak energy preference for the d^0 geometry, then the *trans* isomer is the one which should be found, as it is experimentally. (c) We reserve comment until part (d). (d) The angular overlap energies are evaluated in **3.41**. For simplicity we have used the parameter for metal-oxygen π bonding and ignored the influence of the *CN*

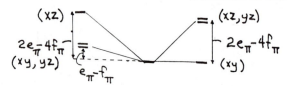

3.41

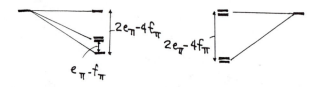

ligands. The coefficient of the fourth order term (which is an essential part of the problem) is simply the square of the second order one. For the d^2 configuration the *trans* form is preferred over the *cis* ($8e_\pi$ - $16f_\pi$ compared to $6e_\pi$ - $10f_\pi$) and the *cis* over the *trans* form for the d^0 configuration ($8e_\pi$ - $12f_\pi$ compared to $8e_\pi$ - $16f_\pi$). Recall $e_\pi, f_\pi > 0$ and $e_\pi > f_\pi$. A similar result comes from numerical evaluation of a secular determinant (discussed more fully in Question 3.23). The deepest-lying orbitals of the *cis* and *trans* geometries, with two metal-ligand interactions, are stabilized less than twice as much as the middle orbitals of the *cis* geometry, with one metal-ligand interaction. (e) The rule is simply that for a molecule where all the bonding orbitals are filled, the most stable geometry for a set of ligands is the one where they share as small a number of central atom orbitals as possible. See the answer to question 3.17 too. (f) The staggered arrangement should be the favored one when all of the relevant metal-ligand bonding orbitals are filled.

3.19. The ligand field strengths from the spectrochemical series are in the order $CN^- > NH_3 > H_2O$ and thus the chemical shifts are expected to increase in this order. Since the chemical shift for the compound containing the ligand L^- is larger than that for water this implies that its ligand field strength should be smaller than that for water too. Small ligand field splittings often give rise to high-spin, paramagnetic systems and indeed this is the case here. In fact the magnetic behavior of these cobalt complexes may be varied by changing the identity of the of the R group in the ligand L^-.

3.20. The orbital diagram is a particularly simple one to derive. If x is the internuclear Cu-Cu axis, then the in-phase xy orbital combination will interact with the p_y orbitals of the bridging oxygen atoms and the out-of-phase combination with the p_x orbitals as shown in **3.42**. The angular variation in the corresponding overlap integrals is shown in **3.43**, and is easy to understand. When $\theta = 180°$ (actually an impossibility here in the real complex) then we can see that the overlap $<xy/y>$ is purely π in nature. As θ decreases then a more eneregetic σ component becomes possible and the overlap increases. The converse is true for $<xy/x>$. Here the smallest overlap should

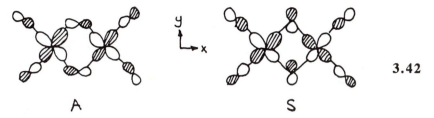

3.42

A   S

occur for $\theta = 0°$. The two should cross at $\theta = 90°$. The variation in energy of these two orbitals with θ (**3.44**) looks very similar to the variation in overlap picture, since they are both *Cu-O* antibonding orbitals. Notice that the energy gap (Δ) at 105° is indeed larger than that at 96°. Although we cannot predict

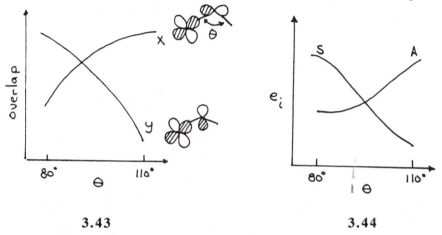

3.43   **3.44**

the exact angle for the crossover from ferromagnetic to ferromagnetic, the trend is certainly in agreement with experiment. (For more details of this classic problem see: *J. Amer. Chem. Soc.*, **97**, 4884 (1975).)

3.21. **3.39** shows the correlation diagram linking square planar and tetrahedral geometries. Complexes of Pt^{II} are always low spin and square planar as is clear from an inspection of this diagram. However, on excitation an electron is forced to occupy a very high energy orbital ($x^2 - y^2$) at this geometry. A rearrangement is then expected towards the tetrahedral geometry in the excited state. (In fact a distorted tetrahedron is probably preferred since such a geometry is Jahn-Teller unstable.) However such a geometry can smoothly relax back to the electronic ground state as shown in **3.45**. Compare this result with that obtained in Question 3.9.

3.22. The orbitals of H_8 are very easy to derive. Take the orbitals of H_6 (these have exactly the same form as those of benzene in Question 6.2) and add a capping "H_2" unit. The deepest lying orbital of H_6 (a_{1g}) interacts with σ_g^+ but all other orbitals, including σ_u^+ on H_2, remain nonbonding . This set of orbitals is used in **3.46** to generate the molecular orbital diagram of MH_8. All the molecular orbitals from $\psi(a_{1g})$ to $\psi(e_{2g})$ are bonding and should be filled.

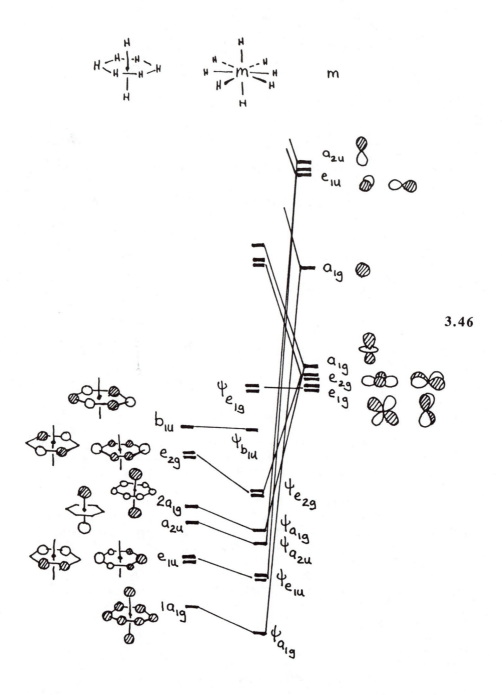

3.46

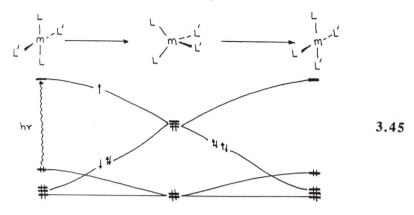

3.45

$\psi(b_{1u})$ is nonbonding. If it is left empty then fourteen electrons are needed. Since the H_8 unit contributes eight electrons, M must contribute six. Therefore, Cr, Mo or W are likely candidates. Provided that the electronegativity of M is considerably less than that of H, we might think that filling $\psi(b_{1u})$ might be acceptable. This then would mean that Fe, Ru or Os would be fine. Filling $\psi(e_{1g})$ requires a Zn, Cd or Hg compound. That will not be too likely. The d orbitals on them are very contracted in this part of the transition metal series and, therefore, $\psi(a_{1g})$ to $\psi(e_{2g})$ on H_8 will not be stabilized by much.

3.23. The orbital problem is set up in **3.47** for the case where the L ligands are π-donors, although the mathematical problem is similar for the acceptor case. We are interested in examining the interaction of the antisymmetric

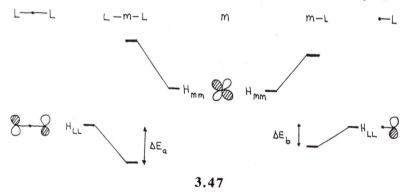

3.47

component $\psi_1 = (1/\sqrt{2})(\phi_1 - \phi_2)$ with the corresponding d orbital, and to compare it with that found from a similar interaction but involving only one ligand orbital, $\psi_2 = \phi_1$. The secular determinant is

$$\begin{vmatrix} H_{MM} - E & H' \\ H' & H_{LL} - E \end{vmatrix} = 0$$

where H' is the interaction integral $<\phi_d|\mathcal{H}^{eff}|\psi_i>$. $|H_{MM}| < |H_{LL}|$ and the lowest lying orbital is largely metal located and M-L bonding. We will use the same parameter for both situations and specify its nature later. Expansion of the secular determinant leads to a quadratic equation in E,

$$E^2 - E(H_{MM} + H_{LL}) - H'^2 + H_{MM}H_{LL} = 0$$

whose roots may be written as:

$$2E = (H_{MM}+H_{LL}) \pm (H_{LL}-H_{MM})\sqrt{(1 + 4H'^2/(H_{LL}-H_{MM})^2)}$$

Expanding under the square root:

$$2E = (H_{MM}+H_{LL}) \pm (H_{LL}-H_{MM})(1 + 2H'^2/(H_{LL}-H_{MM})^2 - 4H'^4/(H_{LL}-H_{MM})^4 +...)$$

which leads to a stabilization energy of the bonding orbital of

$$\Delta E = H'^2/(H_{LL}-H_{MM}) - 2H'^4/(H_{LL}-H_{MM})^3 +...$$

Now we need to evaluate H'^2 and H'^4 for the two cases by specifying the nature of H'. For $\psi_1 = (1/\sqrt{2})(\phi_1 - \phi_2)$, $<\phi_d|\mathcal{H}^{eff}|\psi_i> = (2/\sqrt{2})<\phi_d|\mathcal{H}^{eff}|\phi_1> = \sqrt{2}H_{ML}$, so that

$$\Delta E_a = 2H_{ML}^2/(H_{LL}-H_{MM}) - 8H_{ML}^4/(H_{LL}-H_{MM})^3 +...$$

Thus the stabilization energy *per linkage* is:

$$\Delta\varepsilon_a = \Delta E_a/2 = H_{ML}^2/(H_{LL}-H_{MM}) - 4H_{ML}^4/(H_{LL}-H_{MM})^3 +...$$

For $\psi_2 = \phi_1$, $<\phi_d|\mathcal{H}^{eff}|\psi_i> = <\phi_d|\mathcal{H}^{eff}|\phi_1> = H_{ML}$, so that with only one M-L interaction

$$\Delta\varepsilon_b = \Delta E_b = H_{ML}^2/(H_{LL}-H_{MM}) - 2H_{ML}^4/(H_{LL}-H_{MM})^3 +...$$

It can readily be seen how the stabilization energy can be written as a series of terms which alternate in sign. We expect, since this treatment is very much akin to a perturbation theoretic approach, that the magnitude of consecutive terms decreases. The first term is stabilizing ($H_{LL} < H_{MM}$) but the second (with its minus sign) destabilizing. $\Delta\varepsilon_a$ and $\Delta\varepsilon_b$ contain an identical lead term and only differ in the second. Clearly $|\Delta\varepsilon_a| < |\Delta\varepsilon_b|$ because of the larger destabilizing effect in the case of the two *trans* ligands. (The mathematical discussion here actually forms the basis for the angular overlap model, the system dependent parts of the second order and fourth order terms being labeled e_σ and f_σ respectively. See reference 2.) If bond strength and bond length are inversely related then we can see how such an argument leads in principle to the *trans* influence described in the question. To show this in more detail the argument has be extended to that of two different *trans* ligands, L and L'. The stabilization energy now becomes more complex. Setting $H_{LL} = H_{L'L'}$ as a simplification then

$$\Delta E_a = H_{ML}^2/(H_{LL}\text{-}H_{MM}) + H_{ML'}^2/(H_{LL}\text{-}H_{MM}) - 2H_{ML}^4/(H_{LL}\text{-}H_{MM})^3 -$$

$$2H_{ML'}^4/(H_{LL}\text{-}H_{MM})^3 - 4H_{ML}^2H_{ML'}^2/(H_{LL}\text{-}H_{MM})^3 + ...$$

Dividing the cross-term equally between the two interactions leads to

$$\Delta\varepsilon_a = H_{ML}^2/(H_{LL}\text{-}H_{MM}) - 2H_{ML}^4/(H_{LL}\text{-}H_{MM})^3 - 2H_{ML}^2H_{ML'}^2/(H_{LL}\text{-}H_{MM})^3 + ...$$

and

$$\Delta\varepsilon_a' = H_{ML'}^2/(H_{LL}\text{-}H_{MM}) - 2H_{ML'}^4/(H_{LL}\text{-}H_{MM})^3 - 2H_{ML}^2H_{ML'}^2/(H_{LL}\text{-}H_{MM})^3 + ...$$

Comparison with $\Delta\varepsilon_a$ clearly shows the influence of the *trans* ligand.

3.24. The labels appear to be those appropriate for the O_h point group but with a splitting pattern appropriate for the tetrahedron. In fact the ML_8 cube fits both criteria. It can be regarded as being constructed from two interpenetrating tetrahedra (open and closed circles of **3.48**). From either crystal field or angular overlap models we expect that $\Delta_{cub} = 2\Delta_{tet}$, i.e., $\Delta_{cub} = (8/9)\Delta_{oct}$.

3.48

3.25. Before providing an 'answer' to this problem we must note that the experimental evidence for the existence of such a phenomenon is on shaky ground (see *J. Amer. Chem. Soc.*, **113**, 1437 (1991) and *Inorg. Chem.* **30**, 4433 (1991).) However the electronic picture is so striking that we include the question here.

The orbital correlation diagram is shown in **3.49**. For W^{5+} there should be a single electron in the metal d orbital set. At the left hand side of the diagram there is formally a $W\equiv O$ triple bond since the odd electron resides in a nonbonding orbital. As the W-O bond length increases the xz and yz orbitals will drop in energy since they are W-O antibonding, and for large enough r they will drop below the energy of xy, a purely nonbonding orbital. Now the W-O bond order has been reduced. A schematic energy diagram for the two electronic states that result is shown in **3.50**. The two curves have minima located at different values of r, reflecting the different bond orders for the two states. Notice that along this bond stretching path the one electron in xy must

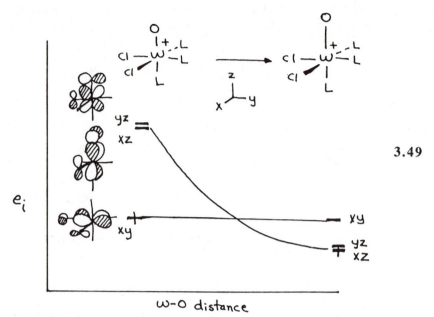

3.49

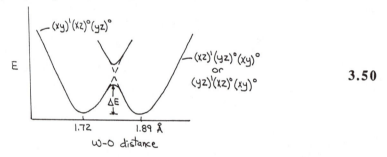

suddenly "jump" to a member of the xz/yz set, a symmetry forbidden reaction. Since the relevant orbitals are orthogonal to each other, there is very little mixing which is allowed between the electronic states, and the activation barrier, ΔE, could be substantial.

3.50

(b) In the first molecule we have suggested a $W\equiv O$ triple bond, which then is partially broken upon going to the second molecule where one $W\text{-}O$ π^* orbital is singly occupied. Notice that the reverse happens to the $W\text{-}Cl$, xy, component. Thus, the $W\text{-}Cl$ distances should decrease as the $W\text{-}O$ bond is stretched.

3.26. One stretching motion of e_g symmetry, in fact the one which leads to either the two long/four short geometry or its converse, is shown in **3.51**. Algebraically it may be written as $\Delta r(e_g) = (1/\sqrt{12})(2\Delta r_1 + 2\Delta r_2 - \Delta r_3 - \Delta r_4 - \Delta r_5 - \Delta r_6)$. If the change in e_σ is proportional to the bond length change then

$$\Delta\varepsilon(z^2) = (1/\sqrt{12})(2\Delta e_\sigma + 2\Delta e_\sigma - (\Delta e_\sigma + \Delta e_\sigma + \Delta e_\sigma + \Delta e_\sigma)/4)$$
$$= (1/\sqrt{12})(3\Delta e_\sigma)$$

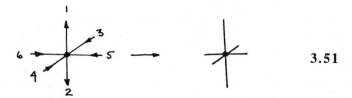

3.51

and

$$\Delta\varepsilon(x^2 - y^2) = (1/\sqrt{12})(0 + 0 - (\Delta e_\sigma + \Delta e_\sigma + \Delta e_\sigma + \Delta e_\sigma)(3/4))$$
$$= (1/\sqrt{12})(-3\Delta e_\sigma)$$

Thus for the two long/four short geometry, Δe_σ as defined, is negative and so z^2 lies lower in energy as in **3.52**. For the four long/two short geometry $x^2 - y^2$ lies lower in energy. Inclusion of the $(n+1)s$ orbital on the metal changes the picture substantially. In the O_h point group this orbital is of different symmetry (a_{1g}) to the metal d orbitals $(e_g + t_{2g})$. However in the distorted

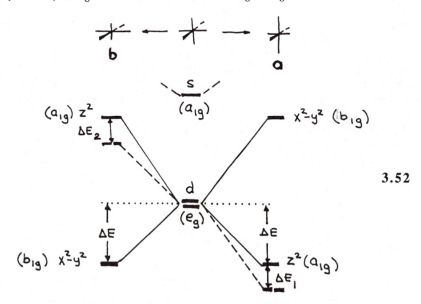

3.52

(D_{4h}) molecule both s and z^2 transform as a_{1g} and may thus mix together. The situation for the two distortions is shown in **3.52** where the dashed lines indicate the orbital energies as a result of this mixing. $|\Delta E_2| > |\Delta E_1|$ on energy gap grounds.

Table 3.5. Stabilization energies on Distortion as a Result of d-s Mixing.

	a	b
d^7	$\Delta E + \Delta E_1$	ΔE
d^8	$2(\Delta E + \Delta E_1)$	$2\Delta E$
d^9	$\Delta E + 2\Delta E_1$	$\Delta E + \Delta E_2$
d^{10}	$2\Delta E_1$	$2\Delta E_2$

For d^7, d^8 and d^9, we can see that distortion **a** is favored, the total stabilization energy consisting of a first-order Jahn-Teller term (ΔE) and a second order one (ΔE_1 or ΔE_2). Notice the largest stabilization for d^8. In fact the square planar complexes universally found for low-spin d^8 complexes may be imagined as being derived from an octahedral parent by a strong Jahn-Teller distortion. For the d^{10} configuration there is only a second order Jahn-Teller contribution which favors distortion **b**. We should expect the largest stabilization of this with the heavier d^{10} systems where the s/d separation is smallest, as in the chemistry of Hg^{II}. The situation is in fact a little more complex than this. Only for relatively small distortions of Cu^{II} complexes is the ratio of axial to equatorial bond length changes close to 2. Invariably the axial distances are much longer. This means that, assuming e_σ scales with bond length, the d orbital stabilization of z^2 in **a** is not equal to the destabilization of $x^2 - y^2$. For the d^{10} system a similar result implies that until the second-order interaction is switched on the distortion **b** may be unfavorable if z^2 goes up in energy more than $x^2 - y^2$ goes down in energy.

3.27. It is important to realize first of all that the collection of filled $M\text{-}L$ σ bonding orbitals prefer a tetrahedral environment over a square planar one. Thus $TiCl_4$, d^0, is a tetrahedral molecule. The answer thus has to lie with occupation of the deepest-lying set of d orbitals, namely those involved in metal-ligand π interactions. What is needed is the addition to a σ only Walsh diagram for the interconversion, the π interactions of the correct symmetry. **3.53** shows those π-donor functions for $X = NR_2^-$ and O^{2-} which find a symmetry match with the metal d orbitals in the two geometries. Notice that at C_{2v} there is less overlap with xy and yz than there is at the D_{2h} geometry. The orbital diagram is thus perturbed as in **3.54**.

matches xy b_{1g}

matches yz b_{3g}

a_2 matches xy

b_1 matches yz

a_1 matches z^2

3.53

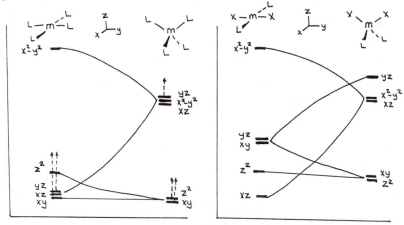

3.54

For $Os(PR_3)_2(NR')_2$ there are four d electrons (Os^{4+}). At the square planar geometry the d orbital configuration is $(xz)^2(z^2)^2$. At the tetrahedral geometry it is $(z^2)^2(xy)^2$. The lower energy of xz, nonbonding as far as the X ligands are concerned, ensures a square planar geometry here. $ReO_2R_2^-$ has a d^2 electron count and on the same same model should also have a square planar geometry. Clearly here though the π forces are not strong enough to overcome the σ preference of the tetrahedron, and the actual geometry lies between the two extremes. For $W(OR)_4$ there is one conformation containing in-plane R groups which leads to the preservation of a π nonbonding xy orbital (**3.55**).

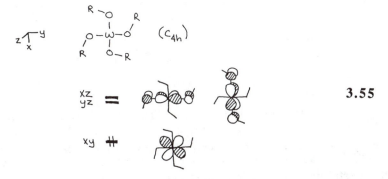

3.55

3.28. **3.39** shows molecular orbital diagrams for the two geometries. The structural result is easy to understand. For the square planar geometry either one or two electrons occupy the weakly antibonding z^2 orbital, whereas in the tetrahedral geometry three or four electrons occupy orbitals which are more strongly σ antibonding. In quantitative terms Table 3.6 shows the *MOSE* values for the two. The stabilization is greatest for the square planar structure, which therefore has the shorter distances. As shown in Question 3.16. the difference between the two is expected to be even larger if d/s mixing is included.

Table 3.6 *MOSE* values for d^7 and d^8 square planar and tetrahedral complexes (Units of e_σ).

	d^7	d^8
square planar (low-spin)	5	4
tetrahedral (high-spin)	4	8/3

3.29. Going across the periodic table from left to right for this series, the metal atom becomes more electronegative and its orbitals consequently become more contracted. Therefore, the *M-O* bond distance is expected to decrease in the order $V^{2+} > Mn^{2+} > Fe^{2+} > Ni^{2+}$. This in fact does happen for the last three members, however, the *V-O* distance is shorter than the *Mn-O* one. Putting electrons in the σ antibonding pair of e_g orbitals will cause the *M-O* bonds to become longer, while population of t_{2g} will not affect the *M-O* distance quite so much. Thus, the *M-O* distances in Mn^{2+}, Fe^{2+}, and Ni^{2+} are simply a reflection of the increasing electronegativity, while the abnormal shortness of the *V-O* distance is a reflection of the fact that there are no electrons in e_g for $V(H_2O)_6^{2+}$. (See Question 3.4 too.) Thus superimposed on a sloping background on moving across the transition metal series (*e.g.*, d^0 to d^3 and d^5 to d^8) are jumps to longer *M-O* distance whenever an electron occupies the e_g orbital **(3.56)**.

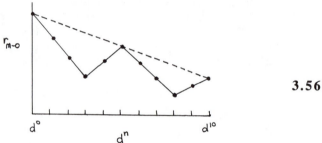

3.56

3.30. An idealized structure is shown in **3.13**. The deoxy species contains iron as Fe^{2+} since the porphyrin carries a 2- charge. Since O_2 contains two electrons in its π^* orbital, there are then eight electrons to be considered when using the Walsh diagram of **3.57**. Such a low spin, eight electron system would have the electronic configuration $(yz+\pi_g)^2(xz+\pi_g)^2(xy)^2(z^2-\sigma_g)^2$. Thus the angle α should be smaller than 180°. The initially formed complex has a d^6 high spin Fe^{2+} unit *plus* the π_g set each singly occupied leading to a configuration $...(\pi_g\text{-}yz)^1(\pi_g\text{-}xz)^1$. As a result the angle α should be ~180°.

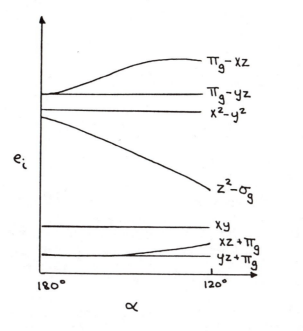

3.57

e_i

α

$\pi_g - xz$

$\pi_g - yz$

$x^2 - y^2$

$z^2 - \sigma_g$

xy

$xz + \pi_g$

$yz + \pi_g$

$180°$

$120°$

Chapter IV.

Main Group Chemistry

*4.1. Construct a molecular orbital argument to rationalise the observed bond lengths in the XeF_n series; XeF_6 (gas, electron diffraction), 1.890 ± 0.005Å; XeF_4 (crystal, neutron diffraction), 1.951, 1.954 ± 0.002Å; XeF_2 (gas, infrared), 1.977 ± 0.002Å. Does your model allow access to similar results for XeF_5^+ and XeF_3^+? In the former the basal bond length is 1.845Å and the 'crossbar' bond length in the latter is 1.89Å.

4.2. The results of electron spin resonance experiments have been used to determine the unpaired central atom spin densities on the central atom for the series of nineteen electron AB_2 radicals, $A = N, P; B = F, Cl$. Interestingly, the spin density observed experimentally increases as the electronegativity difference between A and B increases. At first sight this observation runs counter to established rules. Electron density lies preferentially on the more electronegative atom. By study of the orbital holding the unpaired electron provide an explanation for the observation.

4.3. Molecular and atomic properties often change monotonically on moving down a column of the Periodic Table. The bond energy of F_2 is, however, an anomaly. It is much weaker than expected by extrapolation from the dissociation energies of the higher members. First draw out the molecular orbital diagram and see if you can find a molecular orbital explanation for this. (The observation is a very important one. The stability of many fluorides relies in part upon the low bond energy in this molecule.)

4.4. Use Wade's rules to determine the structures of (a) $C_4(^tBu)_4$ (b) $C_4H_4^{2-}$ (c) P_4 (d) $C_2B_4H_6$ (e) $(CMe)_2(COH)_2(BH)_2$ (a hypothetical molecule, although a transition metal analog does exist).

4.5. Construct a molecular orbital diagram for the ethane molecule (C_2H_6) from the orbitals of two CH_3 fragments and show that B_2H_6 is Jahn-Teller unstable in this geometry.

4.6. (a) Construct a Walsh diagram connecting linear and bent AH_2 molecules where A is a first row $(Li-Ne)$ element. (b) Make sure your diagram contains the symmetry labels for the molecular orbitals and little 'cartoons' showing the molecular orbital composition in terms of atomic orbitals. (c) What geometries do you predict for the molecules BeH_2 and CH_2? Are they the same as predicted by *VSEPR*? (d) Draw out localized orbital descriptions for the bonds in *BeH₂*. (e) What are the symmetry species of the stretching vibrations in the two molecules? Indicate their *IR* and Raman activity. (f) BH_2 is paramagnetic and the g_{ii} entry in the g tensor contains contributions from spin-orbit coupling given by the expression:

$$\lambda . \int \phi_p L_i \phi_q d\tau / (E_q - E_p).$$

L_i is a operator with the same symmetry properties as $\mathcal{R}_i$, a rotation about the i (= x, y, z) axis. λ is the spin-orbit coupling constant. ϕ_p is the orbital wavefunction of the unpaired electron in the electronic ground state (energy E_p) and ϕ_q the wavefunction of the electron in an excited state (energy E_q). Identify those g tensor components (*i.e.*, which i) for which there are non-zero contributions from this source.

4.7.   Devise an orbital model to rationalize the observation that the $n \rightarrow \pi^*$ excitation in N_2X_2 lies in the vacuum ultraviolet (at around 50,000cm⁻¹) for X = F, but in the near infra-red (at around 12,000cm⁻¹) for X = SiH_3. Ensure that your model can also rationalize the much shorter N-N distance found for X = SiH_3 than for X = F.

4.8.   Construct a molecular orbital diagram for the σ manifold of the XeF_4 molecule using s and p orbitals only on Xe. What is the Xe-F bond order? Now repeat the process with d orbitals on Xe too. Which model is correct? Compare your result with the Paulingesque one using d^2sp^3 hybrid orbitals on the central atom.

4.9.   Use p orbitals only on the central Cl atom and use one σ orbital only on each of the F atoms to construct a molecular orbital diagram for ClF_3. Write out the form of the wavefunctions, show the electron occupancy of the set of levels you derive, and explain why there are two long and one short Cl-F distance in the molecule (as distinct from the converse or three equal distances.)

*4.10. In the N_2O molecule the more electronegative atom (*O*) lies at the end of the molecule, but in Ga_2O (a molecule identified only in the gas phase at low concentrations) the oxygen atom lies in the middle. Use the following model to approach this problem. Construct for the homonuclear triatomic A_3, symmetry adapted orbitals for the π manifold of orbitals. Evaluate the coefficients using the Hückel approximation, and the charges associated with central and terminal atoms for the different electron counts possible, and hence rationalize the difference in stability for the two isomers as a function of electron count.

4.11. Provide a molecular orbital explanation for the observation that whereas $N(CH_3)_3$ is pyramidal but $N(SiH_3)_3$ is planar, both $P(CH_3)_3$ and $P(SiH_3)_3$ are pyramidal.

4.12. Use the molecular orbital diagram for the AH_2 molecule from Question 4.6 to answer the following questions for water. (a) What is the connection between the Lewis structure of water and your diagram. Where are the electron-pair bonds? Where are the 'rabbit-ear' lone pairs? (b) Interpret the photoelectron spectrum of water and that of the isoelectronic Ne atom shown in **2.7**.

4.13‡. Construct a molecular orbital diagram for an A_2 diatomic molecule, where A comes from the row of main group elements Li-Ne. Use a program which will generate the eigenvalues and eigenvectors of a real symmetric matrix to show how the energy of the $2\sigma_g$ orbital is very sensitive to the extent of s/p mixing, *via* the size of the $2s/2p$ energy gap of A. Thus comment on the fact that the C_2 molecule is diamagnetic in its electronic ground state. Ionization energies for atomic orbitals are given in Table 1.2.

*4.14. Use the angular overlap model with p orbitals only on the central atom (the Rundle-Pimentel model) to derive the energy levels of a C_{3v} AH_4 molecule as a function of the axial-basal angle θ. Assuming that the central atom s orbital lies deeper than any of the orbitals you have drawn and is a repository for a pair of electrons, show that for the octet molecule the minimum energy within this distortion coordinate lies at the tetrahedral geometry.

4.15. Use Jahn-Teller ideas to predict the structures of ClH_3 and BH_3^+ by starting off from the molecular orbital diagram for the trigonal planar molecule. Compare your results with the predictions of the Walsh and *VSEPR* models.

4.16. Listed in Table 4.1 are some results from an extended Hückel calculation on PH_5 in a D_{3h} geometry. (a) Draw out the form of the molecular orbitals. (b) Predict which positions more electronegative atoms will prefer and what should happen to the axial versus equatorial bond lengths on the basis of the computed charge and overlap populations. (Note, for this calculation all P-H bond lengths were 1.42Å; the off-diagonal elements in the overlap population matrix need to be multiplied by two in order to get the overlap population.)

*4.17. Use p orbitals only on the central S atom and use one σ orbital only on each of the F atoms to construct a molecular orbital diagram for SF_4. Write out the form of the wavefunctions, show the electron occupancy of the set of levels you derive, and explain why the two 'axial' S-F distances are long and the 'equatorial' ones short (as distinct from the converse or four equal distances.) Using the wavefunctions you have generated calculate the charges on each of the fluorine atoms. Hence predict the ligand locations in the most stable structure of $SF_2(CH_3)_2$.

*4.18. The ionization energies associated with the σ-bonds in the permethylated silanes are given in Table 4.2. If each σ-bond was truly localized between the corresponding pair of silicon atoms then only one peak would have been seen

Table 4.1

ATOM		X	Y	Z
P	1	0.00000	0.00000	0.00000
H	2	1.42000	0.00000	0.00000
H	3	-0.71000	1.22976	0.00000
H	4	-0.71000	-1.22976	0.00000
H	5	0.00000	0.00000	1.42000
H	6	0.00000	0.00000	-1.42000

WAVE FUNCTIONS
MO'S IN COLUMNS, AO'S IN ROWS

	1	2	3	4	5	6	7	8	9
P_1(S)	1.5968	0.0000	0.0000	0.0000	0.0170	0.0000	0.0000	0.0000	0.5930
(X)	0.0000	0.0000	0.0000	1.1449	0.0000	0.0000	-0.5840	0.0000	0.0000
(Y)	0.0000	0.0000	1.1449	0.0000	0.0000	-0.5840	0.0000	0.0000	0.0000
(Z)	0.0000	-1.3205	0.0000	0.0000	0.0000	0.0000	0.0000	0.5592	0.0000
H_2(S)	-0.6555	0.0000	0.0000	-0.9809	0.4028	0.0000	-0.4379	0.0000	0.1786
H_3(S)	-0.6555	0.0000	-0.8495	0.4905	0.4028	-0.3792	0.2189	0.0000	0.1786
H_4(S)	-0.6555	0.0000	0.8495	0.4905	0.4028	0.3792	0.2189	0.0000	0.1786
H_5(S)	-0.6219	0.9559	0.0000	0.0000	-0.6130	0.0000	0.0000	0.3719	0.1795
H_6(S)	-0.6219	-0.9559	0.0000	0.0000	-0.6130	0.0000	0.0000	-0.3719	0.1795

```
ENERGY LEVELS (EV)        NUMBER OF ELECTRONS

E(  1) =     30.67492            0.0000

E(  2) =     12.82267            0.0000

E(  3) =      4.36340            0.0000

E(  4) =      4.36340            0.0000

E(  5) =    -11.16647            2.0000

E(  6) =    -17.65932            2.0000

E(  7) =    -17.65932            2.0000

E(  8) =    -18.06050            2.0000

E(  9) =    -22.42674            2.0000


SUM OF ONE-ELECTRON ENERGIES =    -173.94470358 EV.
```

REDUCED OVERLAP POPULATION MATRIX, ATOM BY ATOM, FOR   10
ELECTRONS

	1	2	3	4	5	6
P 1	2.6938	0.6992	0.6992	0.6992	0.5758	0.5758
H 2	0.6992	0.7718	0.0180	0.0180	-0.0870	-0.0870
H 3	0.6992	0.0180	0.7718	0.0180	-0.0870	-0.0870
H 4	0.6992	0.0180	0.0180	0.7718	-0.0870	-0.0870
H 5	0.5758	-0.0870	-0.0870	-0.0870	1.0926	0.0244
H 6	0.5758	-0.0870	-0.0870	-0.0870	0.0244	1.0926

```
          ATOM          NET CHG.

          P    1         0.68165
          H    2        -0.05239
          H    3        -0.05239
          H    4        -0.05239
          H    5        -0.26224
          H    6        -0.26224
```

in the photoelectron spectrum. One way of regarding the series of peaks seen in the spectrum is to consider that the two electrons in an isolated *Si-Si* bond lie in an orbital with an energy of α_{Si-Si} and that adjacent localized bonds interact with each other *via* a term $\beta_{Si-Si/Si-Si}$. Set up the secular determinants for all of the molecues shown and express the occupied orbital energies in the form $E = \alpha_{Si-Si} + x\beta_{Si-Si/Si-Si}$. Use the experimentally determined ionization potentials to determine the values of α_{Si-Si} and $\beta_{Si-Si/Si-Si}$. (Note: it is possible that not all ionizations are observed and reported in Table 4.2.) From your result predict the width of the 'σ' band of polysilanes. (For comparison the analogous value, $\beta_{C\text{-}C\pi/C\text{-}C\pi}$, for conjugated polyenes is about 0.6eV.)

Table 4.2 Ionization energies of permethylated silanes.

$H_3Si-SiH_3$ $H_3Si-SiH_2-SiH_3$ $H_3Si-SiH_2-SiH_2-SiH_3$

8.69 eV   9.19   8.00
                 8.22   8.78
                                         9.38

$Si(SiH_3)_4$ $H_2Si-SiH_2$ H_2SiSiH_2

8.26   7.80   7.75
                  8.90   8.16
                                         9.18

4.19. Calculations on the molecule of **4.1** show that the planar conformation is more stable than the twisted one. This is contrary to the result for allene where the twisted form is more stable in accord with van't Hoff's rule. Construct energy level diagrams for the two isomers concentrating on the π interaction of the B_2H_2 and C_3H_2 fragments and show why the planar form is more stable in this case.

4.1

*4.20‡. The inversion barriers of AH_3 molecules ($A = N, P$) are quite different. NH_3 is quite 'floppy' but PH_3 much more rigid. There is a molecular orbital explanation for this result.

(a) For planar 8 electron AH_3 molecules, the highest occupied molecular orbital (*HOMO*) has a_2'' symmetry and the lowest unoccupied molecular orbital has a_1' symmetry (**4.34**). Use a symmetry adapted combination of the hydrogen orbitals and neglect overlap in the secular determinant. By setting up the appropriate 2x2 secular determinant for the a_1' symmetry orbitals, get an explicit formula for the energy of the *LUMO* using the following parameters:

$e_{1s} = H_{1s1s}$ = energy of the free hydrogen atom orbitals

$e_{ns} = H_{nsns}$ = energy of the A atom s orbital. ($n = 2$ for C, N, O; $n = 3$ for Si, P, S, etc.)

$H_{ss}' = <1s/\mathcal{H}^{eff}/ns> = H_{1sns}$; This is a hamiltonian matrix element connecting the central atom s orbital and a single hydrogen atom s orbital.

(b) Comparison of NH_3 and PH_3.

Use the formula you obtained in (a) to calculate $\Delta E = E(LUMO) - E(HOMO)$ for NH_3 and PH_3 using the values of e_i from Table 1.2 and setting the value of H_{1sns} via the Wolfsberg-Helmholz formula:

$$H_{ss}' = (1/2)(1.75) (e_{1s} + e_{ns}) S_{ss}'$$

Do the calculation assuming $S_{ss}' = .20$, in both cases. (S_{ss}' is the overlap between the central atom ns orbital and a single hydrogen atom $1s$ orbital.)

(c) It is common to represent the radial part of atomic orbitals using Slater type orbitals ($STOs$): $R_{nl} \sim r^n exp(\zeta_l/r)$ for all l. Clementi and Raimondi (*J. Chem. Phys.*, **38**, 2686 (1963)) found optimal values for the ζ_l exponents for the s and p orbitals in Table 4.3

Table 4.3 Exponents for s and p orbitals

atom	ζ_s	ζ_p
N	1.924	1.917
P	1.881	1.629
As	2.236	1.862

Note that ζ_s and ζ_p become increasingly dissimilar as one goes down a column of the periodic table. (Table 1.2 shows the values found from Slater's rules which do not show this trend.) How does this trend affect the relative radial extension of s vs. p orbitals as we go to heavier elements? Because of the lengthening of A-H bonds it turns out to be reasonable to assume that the overlap integrals between the central atom p orbitals and the hydrogen atom s orbitals are fairly constant through this series. If so, should S_{ss}' in (b) be larger for N or P? Adjust the value of S_{ss}' used in (b) by 25% for P (upward or downward in accord with your answer above) and recalculate ΔE.

(d) Discuss the observed trend in H-A-H angles for AH_3 molecules in the light of your results. We often attach great importance to the electronegativity changes of the central atom in comparing these angles. How do you assess the relative importance of electronegativity differences (which are reflected in changes e_s and e_p) versus changes in the relative "extension" (or "contraction") of the s and p orbitals?

4.21. Show how is possible to create stable A_2 molecules (*i.e.*, with a potential energy minimum as a function of internuclear separation) for the inert gases in their excited states.

*4.22. Use perturbation theory to construct a Walsh diagram linking the $D_{\infty h}$ and D_{2h} structures of an A_2H_2 molecule via a pathway which maintains C_{2h} symmetry (**4.2**). The geometry change is the perturbation. Draw out the

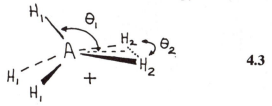

molecular orbitals for the $D_{\infty h}$ and D_{2h} endpoints.

4.23. Consider the $SbCl_5{}^{2-}$ molecule. (a) Determine its structure on the basis of the *VSEPR* rules. (b) Use the Jahn-Teller Theorem to rationalize its distortion away from a D_{3h} geometry to the one predicted in (a). (c) Draw a Walsh diagram for all valence orbitals (neglect the lone pairs on Cl) correlating a D_{3h} geometry with the one you predicted. Briefly explain why each orbital goes up, down or remains at the same energy along the distortion path using perturbation theory.

4.24. Geometry optimization at the *ab initio* level has been carried out for a series of protonated AH_4 molecules. The structure could well be described as that resulting from coordination of H_2 to a pyramidal $AH_3{}^+$ unit (**4.3**).

Explain why θ_1 and θ_2 (Table 4.4) decrease in the order $C > Si > Ge$. Each structure has C_s symmetry.

Table 4.4. Geometry data for protonated AH_4 molecules.

A	C	Si	Ge
$A-H_1$(Å)	1.07	1.45	1.51
$A-H_2$(Å)	1.24	1.94	2.10
θ_1	108.0°	95.4°	94.1°
θ_2	40.4°	22.2°	20.8°

4.25. By allowing the frontier orbitals of two trans AH units to interact with the π orbitals of a square A_4H_4 moeity, show how the octahedral A_6H_6 geometry requires a total of seven skeletal electron pairs for stability (*e.g.*, $B_6H_6{}^{2-}$). Show how the same electron count applies to the *nido* octahedron.

4.26. A number of $(\eta^6 arene)_2Ga^+$ molecules have been made. (a) Draw an orbital interaction diagram for $(\eta^6 benzene)_2Ga^+$ in the D_{6h} geometry shown in **4.4**. Consider only the π and π^* orbitals of the benzene molecule and the valence orbitals of Ga. (b) The geometry of these molecules is however,

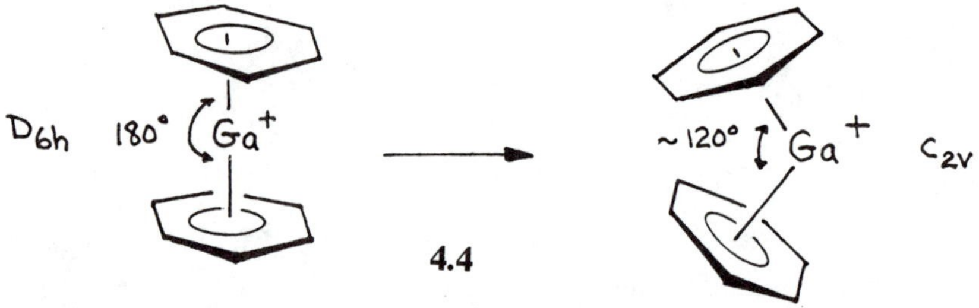

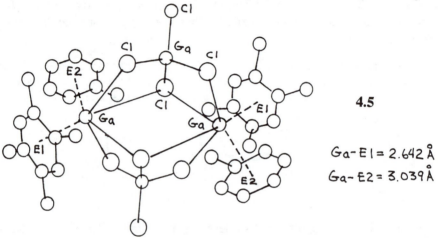

4.4

definitely not D_{6h}, but is 'bent' as shown to a C_{2v} structure. Draw a Walsh diagram for this geometrical change and show by means of simple perturbation theory arguments why the levels go up or down in energy. Hence provide a rationale for the distorted structure. (c) The structure of $[(\eta^6\text{-}C_6H_2Me_4)(\eta^6\text{-}C_6H_5Me)Ga]^+[GaCl_4]^-$ is shown in **4.5** along with some selected bond lengths.

4.5

$$Ga\text{-}E1 = 2.642 \text{ Å}$$
$$Ga\text{-}E2 = 3.039 \text{ Å}$$

The molecule is actually a dimer with two loosely associated $[GaCl_4]^-$ units that 'bridge' the two $(\eta^6 arene)$ units. Notice the $Ga\text{-}E1$ distance (where $E1$ is the center of the $C_6H_2Me_4$ ring) is much shorter than the $Ga\text{-}E2$ distance (where $E2$ is the center of the C_6H_5Me ring). Explain why this occurs.

4.27. For the F_5SCH^{2-} molecule shown in **4.6**, draw out an orbital interaction diagram, using F_5S^+ and CH^{3-}, for only the most important valence orbitals. Draw out the form of the resulting molecular orbitals. (Be careful when you position the initial orbital energies.) There are, by symmetry, two different S-F bonds in the molecule. From your orbital picture decide on their relative lengths.

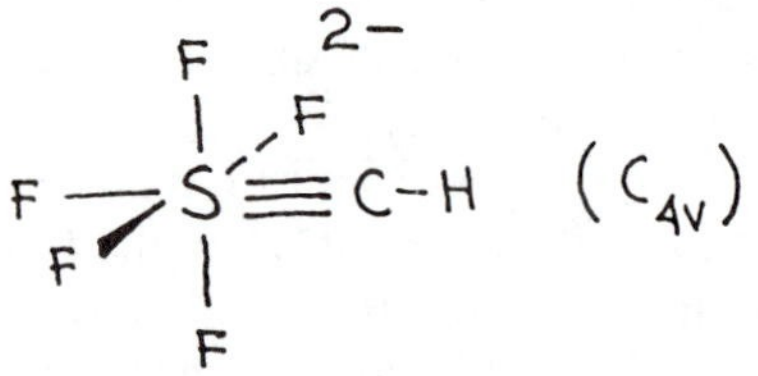

$$F_5S\equiv C\text{-}H \quad (C_{4v}) \qquad \textbf{4.6}$$

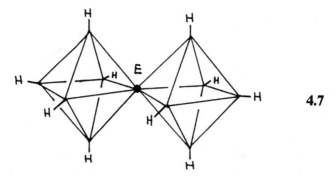

4.7

4.28. Assemble the molecular orbital diagram for the molecule shown in **4.7** from those of two *nido* fragments and a single fusing atom where $E = B$. Hence show that the electron counting rule for fusing two *closo* deltahedra is just the sum of the two electron counts needed for each *closo* molecule. Why does the rule work in practice only for the case where $E = Ga$, and not in fact for when $E = B$?

4.29. Consider the CO_3^{2-} molecule where all O-C-O angles are 120° and all the atoms lie in one plane (**4.8**).

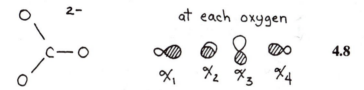

4.8

(a) Form Symmetry Adapted Linear Combinations of the three lone pair orbitals at each oxygen atom. For this purpose, use an *sp* hybrid orbital which points away from the carbon at each oxygen, an atomic *p* orbital which lies perpendicular to the molecular plane at each oxygen and a *p* orbital which lies in the molecular plane (χ_1 -χ_3). Be sure to both draw out each Symmetry Adapted Linear Combination and write out its formula, including the normalization factor. Do the same for the inward pointing hybrids, χ_4. Order each set of Symmetry Adapted Linear Combinations in terms of what you think its relative energy will be.

(b) Now construct a molecular orbital diagram for the system. First switch on the C-O σ interactions followed by the smaller π interactions. Is the carbonate ion a closed shell system on your model?

(c) What is the nominal C-O bond order? Compare your result with that expected from resonance theory.

4.30. By considering the number of skeletal electrons per B-B linkage rationalize the bond length variations shown in **4.9**. Hint: Take into account the coordination numbers of the boron atoms too.

4.31. (a) Derive a molecular orbital diagram for planar AH_3 and explain how and why it changes when the the molecule is distorted to a trigonal pyramid.

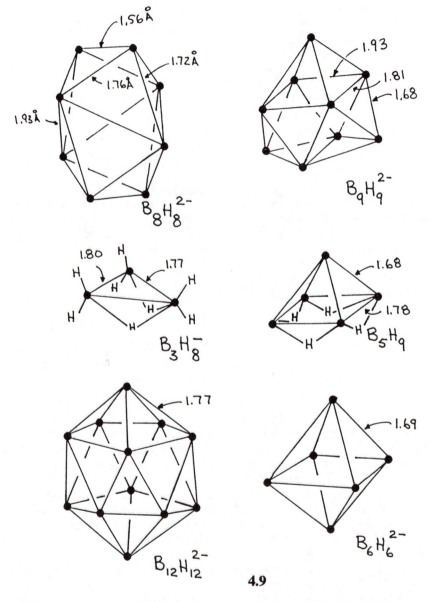

4.9

(b) Use your diagram to rationalize the geometry of NH_3 in its electronic ground state (X) and make a prediction as to its geometry in the first (A) and second (B) excited states which arise from promotion of a single electron from the *HOMO*. Deduce the symmetry species of the states A, B and note whether they can be singlets, triplets or both. (c) For an electronic transition from a state X to a state Y to be dipole allowed the integral $<X/\mu/Y>$ must be non-zero. (μ is the dipole moment operator.) Which of the transitions $X \rightarrow A$ and $X \rightarrow B$ of planar ammonia are dipole allowed and how does the result change for the pyramidal case?

4.32. The torsional angle (**4.10**) in the H_2O_2 molecule is around 90°. Show how *VSEPR* arguments would predict an angle of 180°. Construct a Hückel-type molecular orbital argument to show the origin of this so-called 'gauche' effect. Use the $p\pi$ orbitals of the O_2 unit and the hydrogen $1s$ orbitals. For simplicity set the H_{ii} values of these orbitals equal and also the H_{ij} values describing the interactions between them.

4.33. Use the VSEPR rules to predict the structure for IF_7. Determine the important valence orbitals for the molecule at this structure (disregard the lone pairs on F). Which symmetry inequivalent I-F distances should be longer?

*4.34. The Te_6^{4+} cluster, shown in **4.11**, has a D_{3h} geometry. There are two rather different Te-Te distances in this molecule; $a = 3.13$Å but $b = 2.68$Å. (a) Construct a molecular orbital diagram for Te_6^{4+} from two D_{3h} Te_3^{2+} units and show why $a > b$. (b) Prismane, an isomer of benzene, C_6H_6, has the same basic shape, but here $a = b = 1.55$Å. Explain why the distances are equal in this case.

4.35. Use the *VSEPR* rules to predict the geometries of the molecules I_2Cl_6 and $Hg_2Br_6^{2-}$. Two alternatives are shown in **4.12** for I_2Cl_6.

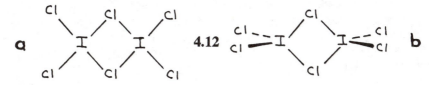

4.36. (a) Construct a Walsh diagram for the distortion of the AH_3 molecule shown in **4.13**. (b) Predict what electron counts are viable ones for either C_{2v} geometry. (c) The optimum geometry for LiH_3 is calculated to be the C_{2v} "T" geometry. Use the Jahn-Teller theorem to show that a distortion from the D_{3h} to this structure is predicted. (d) Predict the reaction path for the rearrangement shown in **4.14**.

H
|
H—A—H

C_{2v}

"T"

H
|
A
/   \
H   H

D_{3h}

H
|
A
/   \
H   H

C_{2v}

"Y"

4.13

$$
\begin{matrix}
\text{H} & & \text{D} \\
| & & | \\
\text{D}-\text{Li}-\text{H} & \longrightarrow & \text{H}-\text{Li}-\text{H}
\end{matrix}
\qquad \textbf{4.14}
$$

4.37. Using a σ-only orbital model for a main-group AX_6 molecule, and ignoring the A atom d orbitals, rationalize the bond-length differences between the following pairs of SbX_6^{n-} molecules.

Table 4.5 Bond-length differences in SbX_6^{n-} molecules.

	$n = 3$	$n = 1$
$X = Cl$	2.65Å	2.35Å
$X = Br$	2.80Å	2.55Å

4.38. Construct a molecular orbital diagram for the molecule B_2H_6 in two steps, B_2H_4 from two BH_2 units and then B_2H_6 by addition of H_2. Comment on the old question as to whether there is a B-B bond in B_2H_6.

4.39. Construct a qualitative π molecular orbital diagram for 'inorganic benzene' $B_3N_3H_6$, where the B and N atoms alternate around the ring. Is the π-electron density located preferentially on B or N?

4.40. Use the molecular orbital diagrams for square planar MX_4 and octahedral MX_6 molecules and add electrons corresponding to the electronic configurations for $M = Xe$ and $X = F$. XeF_4 appears to be a (geometrically) stable molecule, whereas XeF_6 is certainly a 'floppy' molecule with low energy barriers connecting several different structures. From electron diffraction studies one low energy motion which can explain the experimental data is associated with a t_{1u} bending mode. Show how this arises from your molecular orbital orbital diagram. Why doesn't the same thing happen in XeF_4?

4.41. Use molecular orbital arguments to predict the structures of the van der Waals molecules $ArFCl$, $ArCO_2$ and $ArBF_3$.

4.42. Explain the following experimental facts: H-A-H angles: $CH_3^- = 105°$, $NH_3 = 106.7°$, $H_3O^+ = 110°$; inversion barriers: $CH_3^- = 8$, $NH_3 = 5.8$, $H_3O^+ = 2$ kcal /mol.

4.43. (a) Generate the molecular orbital interaction diagram for CO_2. You may assume that the terminal oxygen atoms use sp hybrids for σ bonding, but don't forget to include the nonbonding orbitals built from the lone pair sp hybrids on your diagram. Including the proper orbital polarizations, draw the molecular orbitals for each orbital.

(b) SO_2 and O_3 are bent. Show how the $HOMO$ and $LUMO$ of the linear structure mix upon bending these molecules by using the results from (a). NO_2 is also bent. Should it have a smaller or larger bond angle than O_3?

(c)The singly occupied molecular orbitals ($SOMOs$) for the radicals NO_2

and CO_2^- have been probed using single crystal *ESR* spectroscopy. Measurements of the anisotropic hyperfine and g tensors yield the following (unnormalized) coefficients for the *SOMOs* of these molecules in Table 4.6

Table 4.6. Orbital coefficients for the *SOMO* of NO_2 and CO_2^-.

	2s (on C or N)	2pz (on C or N)	2pz O_1	2pz O_2	2py O_1	2py O_2
CO_2^-	-.374	-.832	.397	.397	~0	~0
NO_2	-.360	-.584	.786	.786	.112	-.112

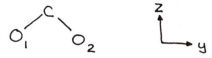

Carefully draw these molecular orbitals. How do they compare with your results from (b)?

(d) The N_2O_4 molecule has a geometry that maximizes the overlap of the *SOMOs* of the NO_2 fragments from which it is built. Draw a simple diagram for the *SOMO-SOMO* interaction for this case. Which should have the stronger central bond, N_2O_4 or B_2F_4?

Answers

4.1. These results show a trend in the opposite direction to that usually observed. Whereas bond lengths tend to increase with increased coordination number, here they decrease. **4.15** shows a set of molecular orbital diagrams for the series XeF_2, XeF_4 and XeF_6, assuming an octahedral geometry for the latter. The deepest-lying occupied orbital is bonding between the Xe 6s orbital and all the ligands, then follow one, two or three bonding orbitals (in XeF_2, XeF_4 and XeF_6 respectively) involving a central atom p_i orbital and the two σ hybrid orbitals, $\phi(j)$ lying along the i axis. These are then followed by two lone pair orbitals in each case. The *HOMO* in each case is the out-of-phase combination of n ligand σ hybrids with the Xe 6s orbital. In general the energetic effects associated with the *HOMO* dominate orbital pictures. This is especially true since the bonding orbitals involving Xe 6p orbitals are of identical form in all three molecules. The *HOMO* may be written as $\psi = a\phi(6s) - b/\sqrt{n}(\Sigma\phi(j))$ where $a, b > 0$ and approximately $a^2 + b^2/n = 1$. The contribution to the bond overlap population of *each* Xe-F linkage from occupation of this orbital is thus proportional to $-4(ab)/\sqrt{n}$. This is controlled by the change in n from 2 to 6, the contribution to Xe-F antibonding decreasing with coordination number. A similar argument applies to the lengths of the three-center bonds in XeF_3^+ and XeF_5.

$2\sigma_u^+$ —

$2e_u$ =

$2t_{1u}$ ≡

$2a_{1g}$ ⧺

$2\sigma_g^+$ ⧺

$2a_{1g}$ ⧺

4.15

π_u ⧻

a_{2u} ⧺

b_{1g} ⧺

e_g ⧻

$1\sigma_u^+$ ⧺

$1e_u$ ⧻

$1t_{1u}$ ⧻

$1\sigma_g^+$ ⧺

$1a_{1g}$ ⧺

$1a_{1g}$ ⧺

4.2. **4.16** shows the resulting orbital character for a situation where two interacting π orbitals have different energies. Notice that the bonding orbital is largely located on the more electronegative atom and the antibonding orbital on the least electronegative atom. This is generally true for all orbital interactions in molecules. Since in stable molecules the number of bonding pairs of electrons exceeds the number of antibonding ones (otherwise the molecule would fall apart) the total electron density predominantly resides on the more electronegative atom. In these nineteen electron AB_2 species the odd electron lies in a π-type orbital of the form shown in **4.17**, antibonding between A and B. Thus its orbital character becomes more A-like as the A/B electronegativity difference increases.

4.3. The molecular orbital diagram for these X_2 diatomics is given in many places including reference 1 page 78 and problem 4.13. Formally, for the Group 17 diatomics the π bond order is zero and the σ bond order equal to 1. The zero π bond order arises because both π and π^* levels are occupied. Recalling Question 2.1, the overall effect of occupation of both bonding and antibonding levels of a set is repulsive. If the overlap integral is significant then overall it can be very repulsive. This is the effect here; the X_2 molecules

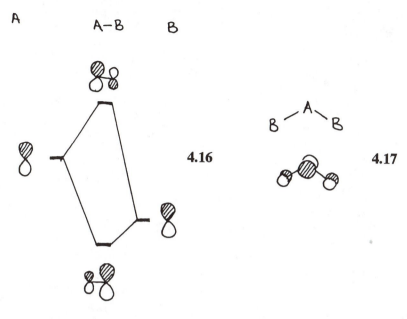

4.16

4.17

are held together by an attractive σ component but a repulsive π component. Recall that for the first row (*Li-Ne*) series π interactions are very important, much more so than for the heavier elements. So the repulsive (π) part of the interaction is much stronger for F_2 than for Cl_2, Br_2 and I_2, giving rise to an abnormally weak *F-F* linkage. Photoelectron data for these diatomic molecules indeed show a large π/π^* splitting for F_2 but not for the heavier analogues. (*Inorg. Chem.*, **22**, 1566 (1983))

4.4. Table 1.5 gives the number of skeletal electrons contributed by each main group fragment. Thus from Wade's rules the structures of these molecules are:
(a) $C_4(^tB_u)_4$ (4x3) = 12 skeletal electrons- *nido* trigonal bipyramid (= tetrahedron).
(b) $C_4H_4^{2-}$ (4x3) +2 = 14 skeletal electrons- *arachno* octahedron (= square plane).
(c) P_4 (4x3) = 12 skeletal electrons- *nido* trigonal bipyramid (= tetrahedron).
(d) $C_2B_4H_6$ (4x3) + (4x2) = 14 skeletal electrons- *closo* octahedron.
(e) $(CMe_2)(COH)_2(BH)_2$ (1x4)+(2x3)+(2x2) = 14 skeletal electrons- *nido* octahedron (= square pyramid).

4.5. The orbitals for ethane are readily constructed from the orbitals of two CH_3 fragments as in **4.18**. As may readily be seen the highest occupied orbital is the antibonding combination of the degenerate (*e* in C_{3v}) $CH_3\pi$ orbitals. Removal of a pair of electrons leads to a degenerate singlet state which is Jahn-Teller unstable.

4.6. (a), (b) This Walsh diagram and its construction is described in many

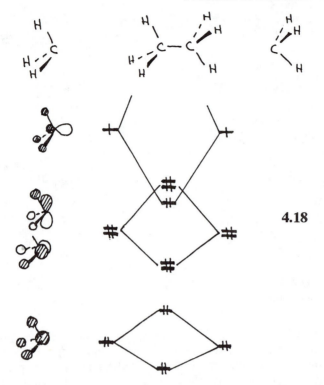

4.18

places including reference 1 page 89 but is shown in **4.19**. (c) BeH_2 is predicted to be linear since the *HOMO* for this species, the $1\sigma_u{}^+$ orbital is destabilized on bending. CH_2 with two more electrons can exist as either a singlet or triplet. As a singlet, with two paired electrons in $\pi_u/2a_1$ its geometry is obviously non-linear. As a triplet with one electron in $2a_1$ and one in b_1 the driving force for bending is reduced. In accord with these ideas the bond angles for singlet and triplet CH_2 are 104° and 133° respectively. For BeH_2 with two pairs of electrons *VSEPR* tells us that the molecule should be linear

4.20

and CH_2 with three pairs should be bent. Presumably *VSEPR* handles the triplet problem by allowing for two 'half-pairs' as in **4.20**.

(d) The localized description of the bonds in BeH_2 are simply described by taking linear combinations of the occupied orbitals. These are shown in **4.21** and show two equivalent *Be-H* bonds.

(e) The symmetry properties of Γ_{str}, the displacement vectors of **4.22** are easy to determine and are shown in Table 4.7. For infra-red activity Γ_{str}

4.21

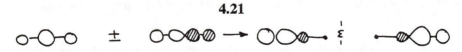

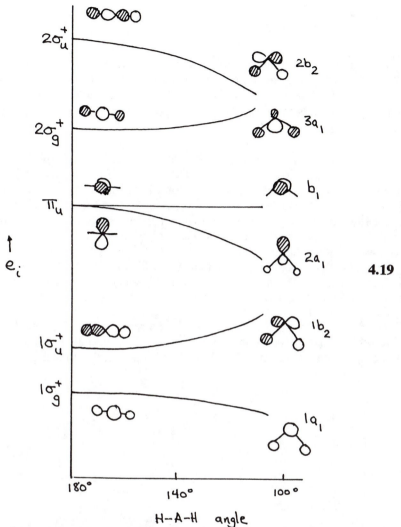

4.19

H–A–H angle

must be contained in Γ_μ (where μ has the same symmetry properties as x, y, z) and for Raman activity Γ_α (where α has the same symmetry properties as xz, xy, x^2 etc.).

Table 4.7. Symmetry properties of Γ_{str}.

	bent (C_{2v})	linear $(D_{\infty h})$
Γ_{str}	$a_1 + b_2$	$\sigma_g^+ + \sigma_u^+$
Γ_μ	$a_1 + b_1 + b_2$	$\sigma_u^+ + \pi_u$
Γ_α	$a_1 + a_2 + b_1 + b_2$	$\sigma_g^+ + \pi_g + \delta_g$

Thus both stretches are *IR* and Raman active in the bent molecule, but for the linear molecule the antisymmetric stretch (σ_u^+) is *IR* and the symmetric stretch

4.22

(σ_g^+) Raman active. The latter result indicates the Rule of Mutual Exclusion for *IR* and Raman activity in centrosymmetric molecules.

(f) The *HOMO* and *LUMO* for *BH$_2$* are of a_1 and b_1 symmetry respectively, and so we need to find the identity of *i* such that $a_1 \times b_1 \times \Gamma_{\mathcal{R}_i} = a_1$. Thus $\Gamma_{\mathcal{R}_i} = b_1$ and from Table 1.1 *i = y*.

4.7. This is just another manifestation of the presence of low-lying π-acceptor orbitals on *SiH$_3$* (see also Question 4.11). As shown in **4.23** the interaction of one component of the *e* σ^* orbital on *SiH$_3$* with the π and π^* levels of the *N$_2$* unit leads to a stabilization of both π and π^*. In *N$_2$F$_2$* where there is a filled deep-lying orbital of this type the corresponding interaction (if any) will be to destabilize both orbitals. Thus the occupied π level is depressed in *(SiH$_3$)$_2$N$_2$* leading to a considerably shortened *N-N* distance (1.17Å) to be compared with that for the *N$_2$* molecule (1.11Å) and alkyl substituted azo compounds (~1.25Å). The π^* level is depressed leading to a low energy $n{\rightarrow}\pi^*$ transition. As a result the molecule is blue.

4.8.   The orbital diagram is shown in **4.15**. The result is three occupied *Xe-F* bonding orbitals (*1a$_{1g}$, 1e$_u$*), one occupied antibonding orbital (*2a$_{1g}$*), and two non-bonding orbitals (*b$_{1g}$, a$_{2u}$*). The *Xe-F* bond order on this model is thus (3-

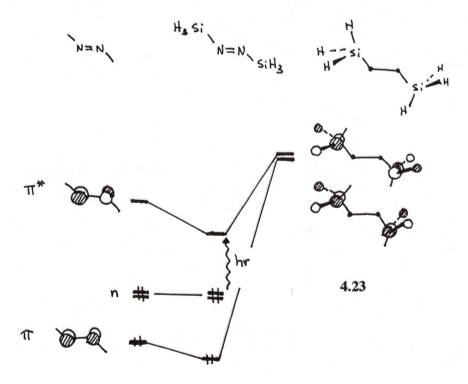

4.23

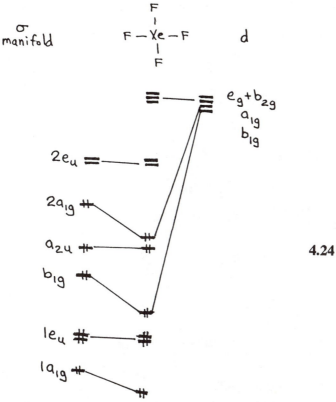

1)/4 = 1/2. The effect of adding in Xe-based d orbitals is shown in **4.24**. The d orbitals transform as $a_{1g} + b_{1g} + b_{2g} + e_g$, and thus stabilize the occupied a_{1g} and b_{1g} orbitals. If the mixing is large enough then $2a_{1g}$ is converted to a nonbonding orbital and b_{1g} is converted to a bonding orbital. Now the Xe-F bond order is (4-0)/4 = 1. By symmetry then two d orbitals get involved in Xe-F bonding, and we can see where the Pauling picture of d^2sp^3 hybrids comes from. Of the two occupied nonbonding orbitals ($2a_{1g}$ and a_{2u}) one is a pure p orbital and the other is a d/s hybrid. Of course the whole question as to which model is correct depends upon the degree of involvement of the d orbitals. Probably the d stabilization here is actually quite small and so the $s + p$ model perhaps more appropriate*.

4.9.   A molecular orbital diagram is assembled for this molecule in **4.25**. In *VSEPR* terms we may regard it as a trigonal bipramidal structure with two equatorial lone pairs of electrons giving a *T* shape. It is immediately apparent that there is one pair of electrons associated with the Cl-F_{eq} linkage, leading to a bond order of one, and only one pair of electrons associated with the two Cl-

*On being asked over lunch one day by one of the authors as to whether d orbitals were important in such species, Robert Mulliken replied "Not really" without taking his eyes of the menu.

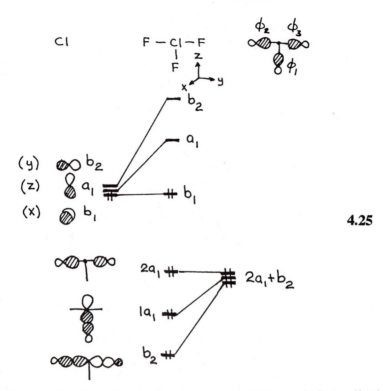

4.25

F_{ax} linkages, leading to a bond order of one half for each. In qualitative terms it is then easy to see how the axial distances should be longer than the equatorial ones. Evaluation of the form of the wavefunctions allows a more quantitative assessment by evaluation of the bond overlap populations. The form of the wavefunctions of the bonding orbitals is just:

$$\psi(1a_1) \sim a\phi(z) + b\phi_1$$

$$\psi(1b_2) \sim a\phi(y) + b(1/\sqrt{2})(\phi_2 - \phi_3)$$

where, ignoring overlap in the normalization process, $b = \sqrt{(1-a^2)}$

Thus the $Cl\text{-}F_{eq}$ overlap population is just $4abS$, where S is the overlap integral of a ligand σ orbital located on a ligand lying along the (say) z axis with the z orbital. Analogously the $Cl\text{-}F_{ax}$ overlap population is just $4ab(1/\sqrt{2})S = 2\sqrt{2}abS$. This difference in overlap population is thus in accord with the result obtained obtained by just counting electrons.

4.10. **4.26** shows the assembly of the π-only diagram for the linear A_3 triatomic. Notice how the orbital coefficients may be simply evaluated within the Hückel approximation since the orbital interactions between the fragments are degenerate ones. (In other words the energies of the ligand combination ψ_1, and the central atom orbital, ϕ_3, are the same.) The electronic charges for two and four pair of electrons in these orbitals (*i.e.*, those appropriate for

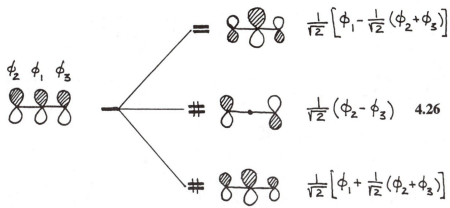

$$\frac{1}{\sqrt{2}}\left[\phi_1 - \frac{1}{\sqrt{2}}(\phi_2 + \phi_3)\right]$$

$$\frac{1}{\sqrt{2}}(\phi_2 - \phi_3) \qquad 4.26$$

$$\frac{1}{\sqrt{2}}\left[\phi_1 + \frac{1}{\sqrt{2}}(\phi_2 + \phi_3)\right]$$

Ga_2O and N_2O) are given in Table 4.8 . Notice that the larger electron density is carried by the central atom in the two pair case, but by the terminal atoms in

Table 4.8   Number of Electrons on the Central and Terminal Atoms of A_3

Atom	Two Pairs	Four Pairs
Central	2.0	2.0
Terminal	1.0	3.0

the four pair case. These then are the sites where, from first order perturbation theory, the more electronegative atom will lie, a result in agreement with experiment.

4.11. This is a similar type of question to that of Question 6.6 where we ask how one can stabilize a planar carbon atom. **4.27** shows how the *HOMO* (a_2'') of a planar AH_3 molecule is of the correct symmetry to interact with ligand π-type orbitals. Notice that if the ligand is a π-donor the result is an energetically unfavorable two orbital-four electron interaction. This may be partially relieved by making the molecule pyramidal. Thus, for species of this type (NF_3, for example), both σ and π effects favor a pyramidal geometry for this

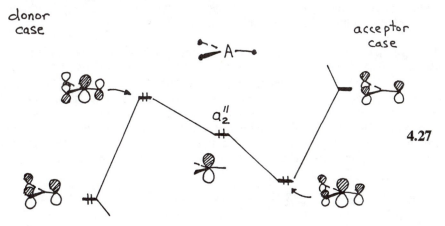

donor case

acceptor case

a_2''

4.27

electron count. For π-acceptor ligands the result is the opposite. A stabilizing interaction results which is maximized where the overlap is largest, namely at the planar geometry. Thus σ effects favor a pyramidal geometry for this electron count, but π effects favor a planar geometry. Which one wins out depends on the system. We may get some estimate of the σ effect on bending by looking at the barriers to inversion of the hydrides, molecules with σ-only ligands. For NH_3 this is around 5 kcal/mole but for PH_3 it is around 35 kcal/mole. This means that it will be much more difficult to stabilize a planar phosphorus atom than a planar nitrogen atom.

For the molecules under consideration the SiH_3 unit has a low-lying $\sigma*$ orbital which is just of the right symmetry to behave as a π-acceptor (**4.23**) and indeed does play a striking stereochemical rôle. It provides an extra stabilization for the planar geometry. In CH_3 this orbital lies much higher in energy. These differences are thus responsible for the planar $N(SiH_3)_3$ molecule but pyramidal $N(CH_3)_3$ species. For the phosphines, the π effect is just not large enough to overcome the preference of the σ manifold for the pyramidal geometry. In several earlier explanations of this effect a low-lying $3d$ orbital (present for Si but absent for C) has often been used in place of a low-lying $\sigma*$ orbital. Recent experimental work has shown that it is the $\sigma*$ orbital which is energetically most important. This is the same orbital which leads to the π-acceptor properties of phosphines, PR_3, which is of vital importance in organometallic chemistry.

A similar effect is important in other areas of chemistry. Geometries at first row atoms are often strongly influenced by the presence of π-acceptor ligands, whereas those at second row atoms are usually dominated by σ effects. In general AX_2 and AX_3 species, where A is a first row atom are either linear or planar, or have a very soft bending distortion. Molecular examples of the latter are found in C_3O_2, $(OC)C(CO)$, and its phosphorous ylid analogue, $(PR_3)C(PR_3)$. Solid state examples center around the structures of the silicates. A very wide range of ...Si-O-Si... angles are found, with very little energetic penalty. As a result the structural diversity of the silicate world is enormous. One important result of this effect is found in the structures of solid oxides and fluorides on the one hand, and sulfides, phosphides and the heavier halides on the other. Framework structures are found for the former (*e.g.*, rutile, TiO_2) which contain planar oxygen or fluorine, but layer structures (*e.g.*, cadmium halide) for the latter which contain pyramidal anions.

4.12 (a) A molecular orbital diagram appropriate for the water molecule is shown in **4.19**. There are two bonding orbitals ($1a_1$, $1b_2$) and two nonbonding orbitals ($2a_1$, b_1), which leads to a ready correlation with the Lewis dot structure containing two bond pairs and two lone pairs. **4.28** shows how localized orbitals may be constructed by taking linear combinations of these two pairs. Notice how the 'rabbit ear orbitals' are generated from such a treatment. It is important to realize that these localized orbitals are not orthogonal, and thus are not stationary states of the Schrödinger wave

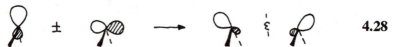

4.28

equation. Thus the 'spectroscopically correct' orbitals are those of **4.19**. (d) This is shown in a very nice way experimentally in the photoelectron spectrum of **2.7**. There are four orbitals shown. The deepest lying is close in energy to an oxygen $2s$ orbital and the other three close in energy to the energy of an oxygen $2p$ orbitals, strong clues as to their orbital parentage. The two bonding orbitals have different energies, as do the two nonbonding orbitals. The latter should be of the same energy if the 'rabbit-ear' model was correct. Notice the smooth correlation with the $2s$ and $2p$ levels of *Ne*. There is now *experimental* evidence as to the atomic orbital composition of these levels (*Chem. Phys.*, **143**, 1 (1990)) which confirms this picture.

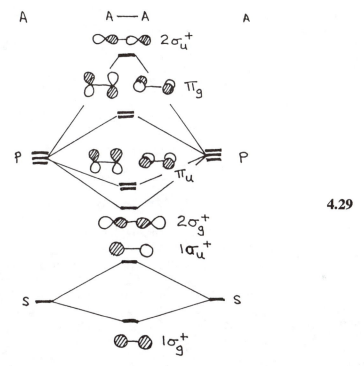

4.29

4.13. This is a classic freshman chemistry problem, but one which is rarely well understood. The best way to see how the diagram comes out is to first of all include only the degenerate orbital interactions as in **4.29**. This includes s-s and p-p interactions but ignores s-p. Now we recognize that there is a σ bonding orbital for the s orbital manifold and one for the p orbital manifold of the same symmetry which can interact (s-p mixing). A similar picture applies to the corresponding σ antibonding orbitals. In **4.30** they are allowed to interact. The deeper-lying member of each pair is pushed down and the higher-lying member pushed up in energy. The new orbitals are s/p hybrids and are readily generated using conventional rules. **4.31** shows how they come about for the orbitals of σ_g^+ symmetry. Perturbation theory tells us that the extent of the mixing is inversely proportional to the energy separation of the mixing orbitals- in this case this is roughly the $2s/2p$ gap. This is larger at the right hand side of the Periodic Table than the left (See Table 1.2), and so such

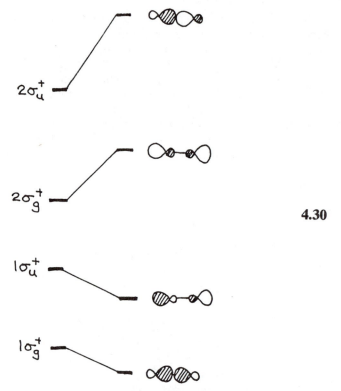

4.30

mixing should be more important for *Li* than for *F*. It is so significant for *C* that the $2\sigma_g{}^+$ orbital lies higher in energy than π_u. As a result the C_2 molecule has a closed shell with a $(\pi_u)^2$ configuration. Numerically we used the ionization energies in Table 1.2 and solved the 4x4 determinant connecting the 2s and 2p orbitals. To keep matters consistent we used the same value of H_{ij} for the three types of interaction. The results are shown in **4.32** and show a shift of around 4.1, 2.1 and 1.1eV for *Li, C* and *F* respectively confirming the result expected from perturbation theory.

4.14. For the C_{3v} AH_4 geometry, the central atom p orbitals transform as a_1 + *e* and the ligand orbitals as $2a_1$ + *e*. A qualitative molecular orbital diagram is shown in **4.33**. Since the overlap integral of a ligand σ orbital with a central atom z orbital is given by $S(\theta) = S_\sigma\cos\theta$, the sum total second order interaction is just $\varepsilon(a_1) = e_\sigma + 3e_\sigma\cos^2\theta$, the first term associated with the single axial interaction and the second with the basal trio. The interaction with each of the *e* symmetry orbitals is then $\varepsilon(e) = (3/2)e_\sigma\sin^2\theta$. This comes from the sum rule since $\varepsilon(a_1) + 2\varepsilon(e) = 4e_\sigma$. The angular dependence of a fourth order term is

4.31

4.32

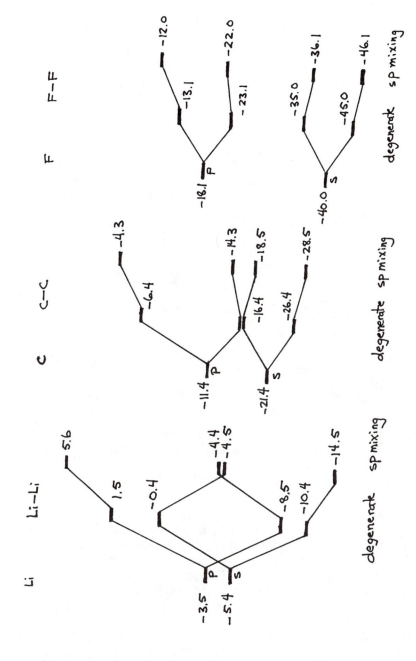

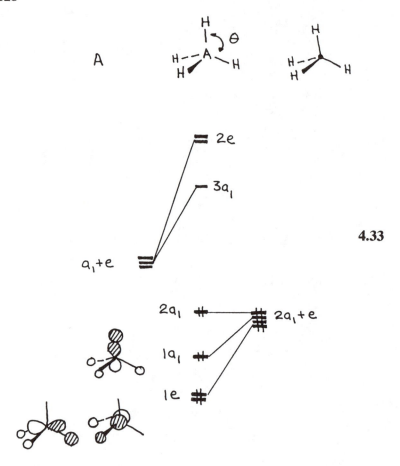

4.33

just the square of that for the corresponding second order one. This leads to a total stabilization energy of the three doubly occupied p orbital components of;

$$8e_\sigma - 2[(1 + 3\cos^2\theta)^2 + 2(3/2)^2\sin^4\theta]f_\sigma$$

The second term is destabilizing but has a minimum value when $\cos\theta = 1/3$ i.e., at the tetrahedral angle. Thus on electronic grounds, using a model which does not include ligand-ligand interactions, the lowest energy geometry is the tetrahedral one. (A similar argument applies to the six-coordinate octahedral structure for twelve electron AH_6.) Contrast this approach to that of *VSEPR* where it is the repulsion between the localized bond pairs which lead to this structure. Steric arguments will always favor the most symmetrical geometry.

4.15. A Walsh diagram for pyramidalizing a trigonal planar AH_3 molecule is shown in **4.34**. The electronic configurations of ClH_3 and BH_3^+ are $(1a_1')^2(1e')^4(a_2'')^2(2a_1')^2$ and $(1a_1')^2(1e')^3$ respectively. BH_3^+ with a ground electronic state of $^2E'$ symmetry is thus first-order Jahn-Teller unstable. From problem 3.12 we know that the Jahn-Teller active mode for a state of this

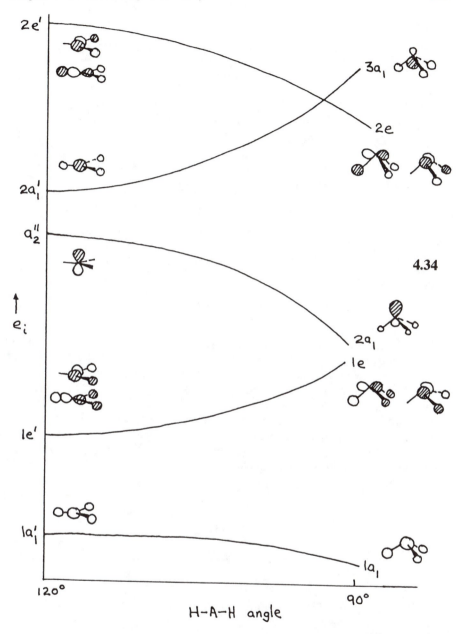

4.34

H–A–H angle

symmetry in this point group is e', which takes the trigonal plane to either a Y-or T-shape (**3.34**). ClH_3 has a ground electronic state of $^1A_1'$ symmetry and is thus first-order Jahn-Teller stable. However, there is a low-lying $^1E'$ state arising from the configuration $(1a_1')^2(1e')^4(a_2'')^2(2a_1')^1(2e')^1$ which can give rise to a second order Jahn-Teller instability associated with a motion of $a_1' \times e' = e'$ symmetry too. In fact ClH_3 is a T-shape and BH_3^+ calculated to be a Y-shape molecule.

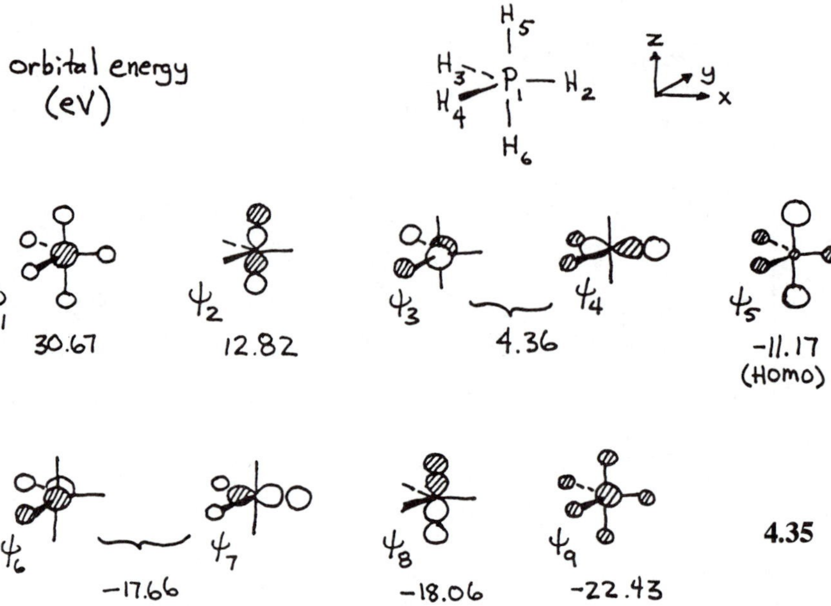

orbital energy
(eV)

ψ_1 30.67 ψ_2 12.82 ψ_3 ψ_4 4.36 ψ_5 -11.17
                                                                              (HOMO)

ψ_6 ψ_7 ψ_8 ψ_9 4.35
        -17.66   -18.06   -22.43

4.16. (a) The orbitals are drawn out in **4.35**. Note that a positive coefficient for (say) a p_x orbital indicates that the positive lobe of this orbital points in the $+x$ direction. We have used shading where the wavefunction is positive. (b) Electronegative substituents should go into positions of highest electron density. Therefore (**4.36**) the axial positions (H_5 and H_6) are preferred over the equatorial ones (H_2, H_3 and H_4). In the calculation both P-H linkages are set at the same distance. Since the axial bonds are predicted by the Mulliken overlap population (**4.37**) to be weaker on this model, in the real molecule they should be longer than the equatorial ones.

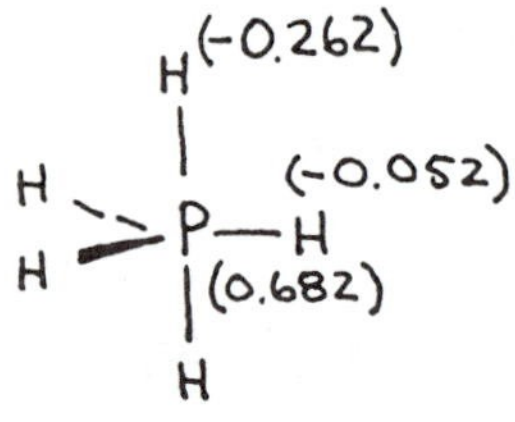

4.36 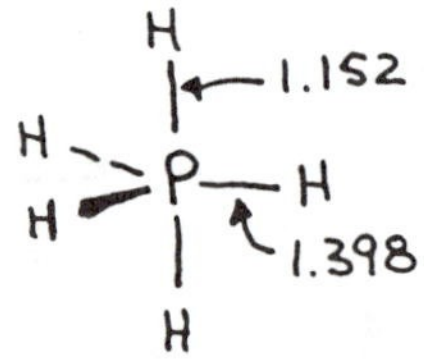   **4.37**

4.17. A molecular orbital diagram is assembled for this molecule in **4.38**. In *VSEPR* terms we may regard it as a trigonal bipramidal structure with one equatorial lone pair of electrons. It is immediately apparent that there are two pairs of electrons associated with the two S-F_{eq} linkages, leading to a bond order of one for each, but only one pair of electrons associated with the two S-F_{ax} linkages, leading to a bond order of one half for each. In qualitative terms it is then easy to see how the axial distances should be longer than the equatorial ones. Evaluation of the form of the wavefunctions allows a more quantitative assessment by evaluation of the bond overlap populations. The form of the wavefunctions is just:

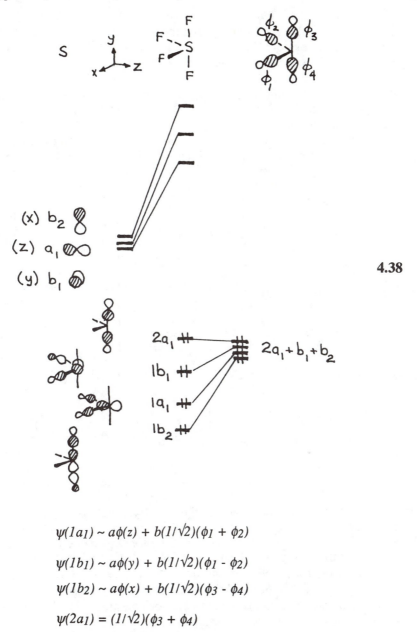

4.38

$$\psi(1a_1) \sim a\phi(z) + b(1/\sqrt{2})(\phi_1 + \phi_2)$$

$$\psi(1b_1) \sim a\phi(y) + b(1/\sqrt{2})(\phi_1 - \phi_2)$$

$$\psi(1b_2) \sim a\phi(x) + b(1/\sqrt{2})(\phi_3 - \phi_4)$$

$$\psi(2a_1) = (1/\sqrt{2})(\phi_3 + \phi_4)$$

where, ignoring overlap in the normalization process, $b = \sqrt{(1-a^2)}$

Thus the $S\text{-}F_{eq}$ overlap population is just $(4)2ab(1/\sqrt{2})S.cos(45°) = 4abS$, where S is the overlap integral of a ligand σ orbital located on a ligand lying along the (say) z axis with the z orbital. Analogously the $S\text{-}F_{ax}$ overlap population is just $4ab(1/\sqrt{2})S = 2\sqrt{2}abS$. This difference in overlap population is thus in accord with the result obtained obtained by just counting electrons. The charge density on F_{eq} is just $2b^2$, but that on F_{ax} is $1 + b^2$. The axial fluorine atoms thus carry the higher electronic charge, and therefore in a substituted sulfurane will attract the more electronegative atoms or groups. In $SF_2(CH_3)_2$ the CH_3 groups are therefore found in the equatorial positions.

4.18. The secular determinants are readily set up. We show two here. Note that the one for the cyclic $(SiH_2)_5$ molecule is *topologically* identical to that for the $p\pi$ orbitals of cyclopentadienyl whose roots are shown in **1.5**. However the *four* silicon open chain molecule $SiH_3SiH_2SiH_2SiH_3$ is equivalent to the *three* carbon allyl unit since there are only three entries (three Si-Si bonds) in the secular determinant.

For $(SiH_2)_5$

$$\begin{vmatrix} \alpha-e & \beta & 0 & 0 & \beta \\ \beta & \alpha-e & \beta & 0 & 0 \\ 0 & \beta & \alpha-e & \beta & 0 \\ 0 & 0 & \beta & \alpha-e & \beta \\ \beta & 0 & 0 & \beta & \alpha-e \end{vmatrix} = 0$$

For $SiH_3SiH_2SiH_2SiH_3$

$$\begin{vmatrix} \alpha-e & \beta & 0 \\ \beta & \alpha-e & \beta \\ 0 & \beta & \alpha-e \end{vmatrix} = 0$$

The secular determinant appropriate for the $Si(SiH_3)_3$ molecule is not given in **1.5** since there is no $p\pi$ equivalent. You will have to solve for it numerically. The roots are $e = \alpha - \beta$ (three times) and $e = \alpha + 3\beta$. Organizing the data given in Table 4.2 with these results leads to the plot of **4.39** taken from *Angew. Chem. Int. Ed.*, **28**, 1627 (1989). From this we can readily determine $\alpha_{Si-Si} =$ 8.7eV and $\beta_{Si-Si/Si-Si} = 0.5eV$. The width of an energy band defined by an interaction parameter β is simply equal to 4β (see Question 8.1). Thus the band width in the polymer is expected to be around 2eV.

4.19. By simple electron counting, the molecule is isoelectronic with allene. Each trigonal planar boron atom forms three bonds, one B-B, one B-H and one B-C. If this were the whole story, by reference to the π levels of the two X_2C-C-CX_2 conformations shown in **6.39**, the molecule should adopt the allene conformation (see Question 6.17). However, consideration of the π orbitals of the B_2H_2 unit changes the picture dramatically. We need to ask in which conformation will the maximum stabilization between the boron located and

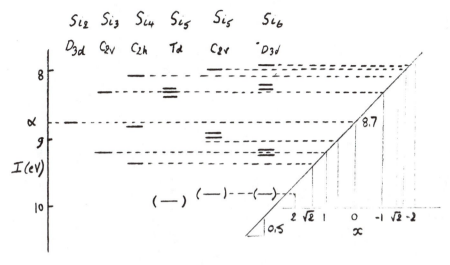

4.39

carbon located π orbitals occur. The two π orbitals form a *B-B* bonding and antibonding orbital pair. The out-of-phase arrangement is of the wrong symmetry to interact with a carbon π orbital, but the in-phase arrangement is nicely set up to interact with the relevant terminal carbon π orbital. As can be seen from **4.40** the largest stabilization occurs in the planar geometry since the nonbonding C_3 orbital is closest in energy to the relevant boron $p\pi$ level. The bonding and antibonding *B-B* π levels have been drawn higher in energy than those on the C_3 unit to relect the difference in electronegativity. If the *B-B* bonding level lay deeper in energy then the result would be formal transfer of two electrons from C_3 to B_2 resulting in a 2 π-electron count for the C_3 unit. As shown in Question 6.17 for this electron count the twisted form is energetically preferred.

4.20. (a) The symmetry adapted hydrogen combination is just $(1/\sqrt{3})(\phi_1 + \phi_2 + \phi_3)$ which can overlap with the central atom ns orbital *via* the integral $<(1/\sqrt{3})(\phi_1 + \phi_2 + \phi_3)/\mathcal{H}^{eff}/\phi_{ns}> = \sqrt{3}H_{ss'}$. The secular determinant is thus

$$\begin{vmatrix} H_{ns} - E & \sqrt{3}H_{ss'} \\ \sqrt{3}H_{ss'} & H_{ss} - E \end{vmatrix} = 0$$

Solution of the determinant leads to

$$E^{\pm} = [(H_{ns} + H_{ss}) \pm [(H_{ns} - H_{ss})^2 + 12H_{ss}{'}^2]^{1/2}]/2$$

The upper root is the *LUMO* of the eight-electron AH_3 molecule.

$$E_{LUMO} = [(H_{ns} + H_{ss}') + [(H_{ns} - H_{ss}')^2 + 12H_{ss}{'}^2]^{1/2}]/2$$

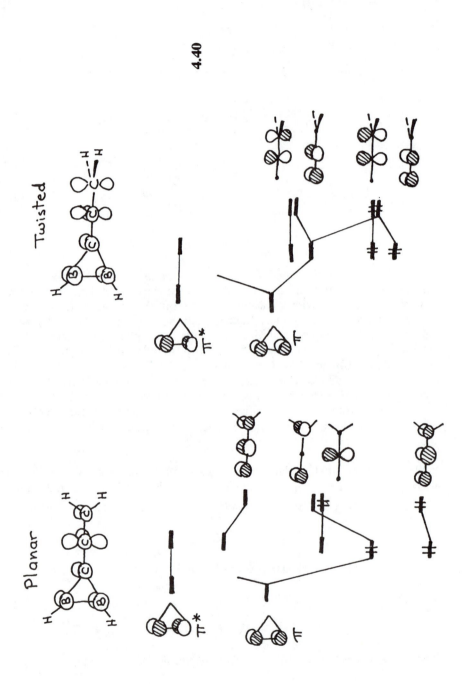

4.40

(b) The *HOMO* is the *np* orbital of the *A* atom, so for *NH₃*

$E(LUMO) = -19.8 + [(12.4)^2 + 12(0.875)^2(39.6)^2(0.2)^2]^{1/2}]/2 = -6.29eV$
$\Delta E = -6.29 + 13.4 = 7.11eV$

For *PH₃*

$E(LUMO) = -16.1 + [(5.0)^2 + 12(0.875)^2(32.2)^2(0.2)^2]^{1/2}]/2 = -6.025eV$
$\Delta E = -6.025 + 12.5 = 6.475eV$

(c) (i) *ns* orbitals become increasingly contracted as one goes down the periodic table. (ii) The relative *s* orbital contraction causes the $S_{ss'}$ overlap to become smaller on moving to heavier elements. is smaller for *PH₃*. (iii)

$$\Delta E = 12.5 - 16.1 + [(5.0)^2 + 12(0.875)^2(32.2)^2(0.15)^2]^{1/2}]/2$$
$$= 4.135eV$$

(iv) The driving force for distortion derives mainly from mixing the *HOMO* and *LUMO*. We know from perturbation theory that this mixing is proportional to $1/\Delta E$. Therefore we expect *PH₃* (and *AsH₃* as well) to bend more than *NH₃* as is indeed observed. (v) The numerical results clearly implicate the s orbital contraction as the more important effect (Compare (b) and (iii).)

4.21. **4.29, 4.30** show molecular orbital diagrams using the valence orbitals (*ns* and *np*) of *A* for the *A₂* diatomic molecule. We know that in the electronic ground states for these inert gas dimers, the number of electrons occupying bonding and antibonding orbitals is equal leading to a bond order of zero. Since antibonding orbitals are destabilized more than their bonding counterparts are stabilized the overall result is a repulsive interaction between the two *A* atoms. **4.41** shows a molecular orbital diagram where the *(n+1)s* orbitals are included too. Promotion of an electron from the highest energy σ* orbital formed from *ns, np* overlap to the lowest energy σ bonding orbital formed from *(n+1)s, (n+1)p* overlap converts an antibonding electron into a

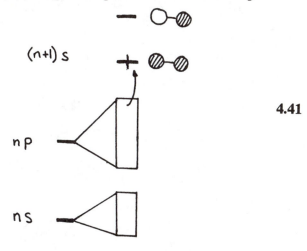

4.41

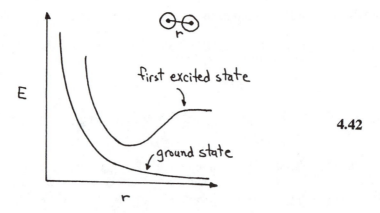

4.42

bonding one. The bond order is now one half. A pair of schematic potential energy curves for the ground and first excited states is shown in **4.42**. A similar result is shown in Question 7.7.

4.22. First develop the orbitals of A_2H_2 in the $D_{\infty h}$ and D_{2h} geometries. **4.43**

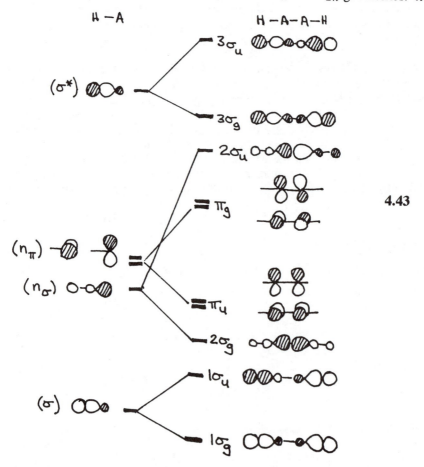

4.43

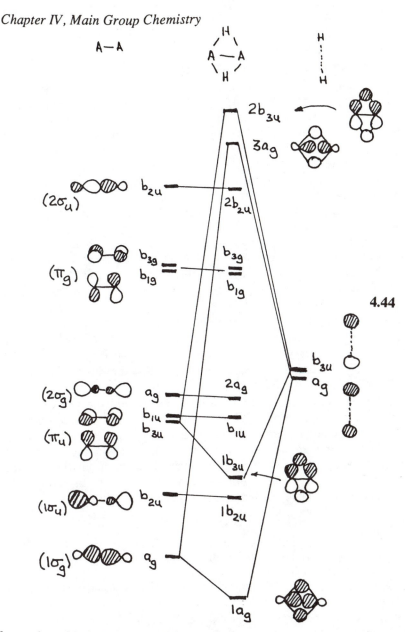

4.44

shows the orbital diagram and the orbital shapes for the $D_{\infty h}$ structure. Notice that the magnitude of the splitting follows $n_\sigma > n_\pi \sim \sigma \sim \sigma^*$. Very little non-degenerate mixing occurs. Listed in parentheses are the symmetry species of each orbital under C_{2h} symmetry. The most convenient way to get the orbitals for the D_{2h} geometry (**4.44**) is to interact A_2 with H_2. Notice that $2\sigma_g$ does not interact very much because of its hybridization.

To construct the Walsh diagram using perturbation theory it is easiest to start from the $D_{\infty h}$ side. Recall that;

$e_i(1) \propto S_{ii}$ and

$e_i(2) \propto \Sigma_{j \neq i} S_{ij}^2 / e_i(0) - e_j(0)$ with $/e_i(1)/ > /e_i(2)/$

The $e_i(1)$ corrections are easy to evaluate. The difficult part is associated with the fact that there are four MO's of a_g and four of b_u symmetry. Their intermixing is quite strong. The resultant Walsh diagram is shown in **4.45**.

The two easiest levels to understand are the a_u component of π_u and the b_g component of π_g. For both the $e_i(1)$ and $e_i(2)$ terms are zero. Let's proceed now from the bottom to the top of the diagram. The $1\sigma_g$ orbital should correlate with $2a_g$ and the $2\sigma_g$ with $1a_g$. They undergo an avoided crossing as shown by the dashed lines. Initially $1\sigma_g$ rises because overlap (leading to a bonding interaction) is lost ($e_i(1) = +$). However the higher energy $2\sigma_g$ mixes into $1\sigma_g$ in a bonding way, ($e_i(2) = -$). Halfway along the distortion coordinate, overlap between the hydrogen AO's and the s/p hybrid on the A atoms is switched back on so that now $e_i(1) = -$. The basic trend in $2\sigma_g$ is the mixing with $1\sigma_g$ ($e_i(2) = +$). The $1\sigma_u$ orbital loses overlap ($e_i(1) = +$), however the second order mixing of the b_u component of π_u keeps it from rising too high in energy. The $e_i(1)$ value for the b_u component of π_u is zero. However, $1\sigma_u$ and $3\sigma_u$ (with the latter dominant) mix in second order. The a_g component of π_g has an interesting behavior. The value of $e_i(1)$ is zero. Initially $3\sigma_g$ mixes strongly into it in second order so that it is stabilized and $3\sigma_g$ rises in energy. However, around the midpoint of the distortion overlap between the two MO's is diminished so that b_{1g} and b_{3g} lie again at the same energy for the D_{2h} structure. The $3\sigma_g$ orbital has the added effect of losing overlap of the antibonding type ($e_i(1) = -$). Thus it does not rise to too high of an energy. Notice that $3a_g$ will lie at a lower energy than $3\sigma_g$. They have approximately the same amount of A-H bonding, but A-A bonding is larger in $3a_g$ than in $3\sigma_g$. Very little happens in $2\sigma_u$ (a small amount of bonding is lost). Finally in $3\sigma_u$ $e(1)$ is large and negative (antibonding interaction is lost), but the b_u component of π_u mixes strongly in second order so that $e(2)$ is large and positive.

4.23. (a) With $5(Sb) + 5(5xCl) + 2$(charge) $= 12$ electrons, the geometry should be based on an octahedron of electron pairs, namely a square pyramid with one lone pair as in **4.46**.

(b) The molecular orbital diagram for the D_{3h} trigonal bipyramid is shown in **4.35** (Question 4.16). The *HOMO* ($3a_1'$) of this electron configuration is close in energy to that of the *LUMO* ($2e'$) leading to a possible Jahn-Teller distortion of species $a_1' x e' = e'$. A motion of this type is shown in **3.19** and takes the trigonal bipyramid to the square pyramid.

(c) A Walsh diagram connecting the two geometries is shown in **4.47**. Since the *Cl-Cl* distance is long we only need to consider the changes associated with central atom-ligand interactions as the geometry is changed. The orbital energy changes may be described as follows:

$1a_1'$: $e(1) \approx 0$, $e(2) \approx 0$ (a very small mixing of the y component of $1e'$.)

$1a_2''$: $e(1) = 0$, $e(2) = 0$ (no change in overlap)

$1e'$ (x component): $e(1) < 0$, $e(2) = 0$ (increased overlap between Sb x and Cl hybrid orbitals)

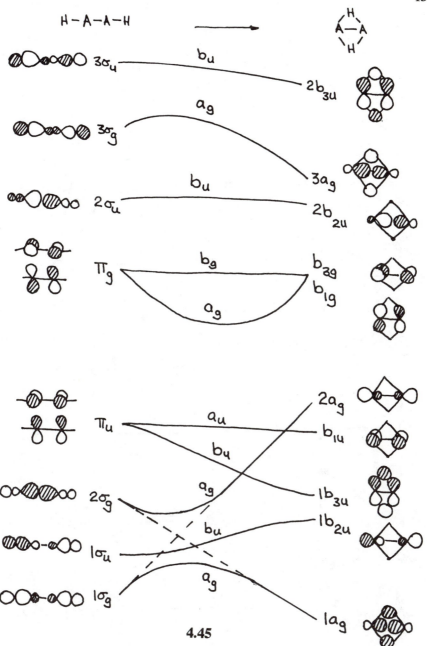

4.45

(y component): $e^{(1)} > 0$ (decreased overlap between *Sb* y and *Cl* hybrid orbitals), $e^{(2)}$ = small (-ve), mixing with $2a_1'$ and even smaller with $3a_1'$.

$2a_1'$: $e^{(1)} \approx 0$, $e^{(2)} \approx 0$ (small mixing with $3a_1'$ which is stabilizing and a small mixing with the y component of $1e'$ which is destabilizing.)

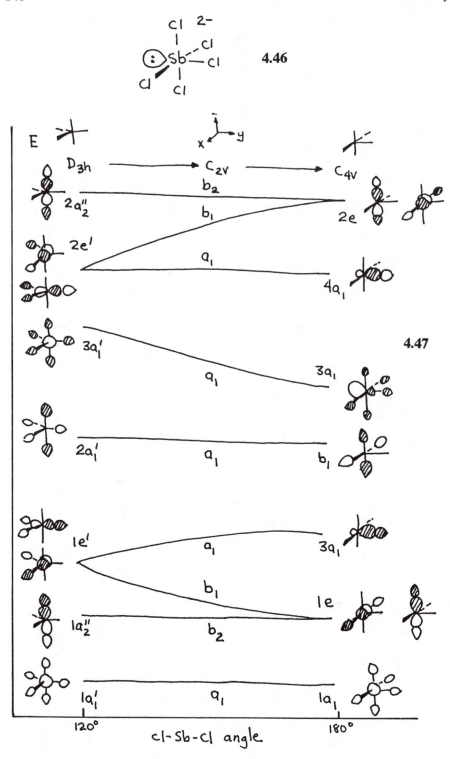

cl-Sb-cl angle

$3a_1'$: $e(1) \approx 0$, $e(2) < 0$ (large mixing with the y component of $2e'$)
$2e'$ (x component): $e(1) > 0$, (increased overlap see $1e'$), $e(2) = 0$.
  (y component): $e(1) < 0$ (decreased see $1e'$), $e(2) > 0$ (large mixing with $3a_1'$.)

4.24. Notice that the $H...H$ distance for $A = Ge$ is quite short (0.76Å) compared to that for $A = C$ (0.86Å). The difference between the two A-H distances though is much greater for $A = Ge$ than it is for $A = C$. This suggests that it will be profitable to study the interaction of the H_2 unit coordinated to an AH_3^+ fragment. An orbital interaction diagram is shown in **4.48**.

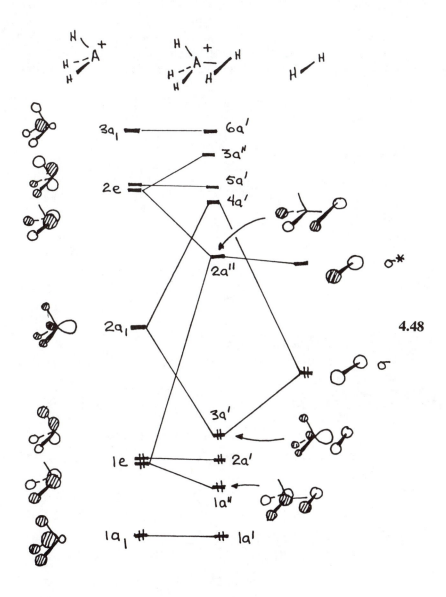

4.48

The critical interactions are twofold: The first is the donation of electron density from the filled H_2 σ orbital into the empty $2a_1$ orbital of the pyramidal fragment. The larger this interaction, the longer the H...H distance. The second is back-donation from one member of the filled $1e$ set into H_2 σ*. Again the stronger the interaction the weaker the H...H bond of the coordinated H_2. On moving from $A = C$ to $A = Si$ we expect that the increase in principal quantum number is accompanied by a reduction in overlap between $2a_1$ and H_2 σ. In addition that between $1e$ and H_2 σ* should decrease too. There is a simple explanation for this. As A becomes less electronegative, the $1e$ set becomes less localized on A (analogously $2e$ becomes more localized on A) and the overlap with H_2 σ* drops.

4.25. **4.49** shows the assembly of the molecular orbital diagram for the octahedron. At left are blocks of orbitals containing the four outward pointing σ A-H bonding and antibonding orbitals, and the in-plane A-A σ bonding orbitals for the square A_4H_4 moeity. Above them are the π orbitals of A_4H_4 whose spatial extent leads to a good interaction with the frontier orbitals (one

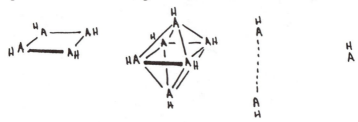

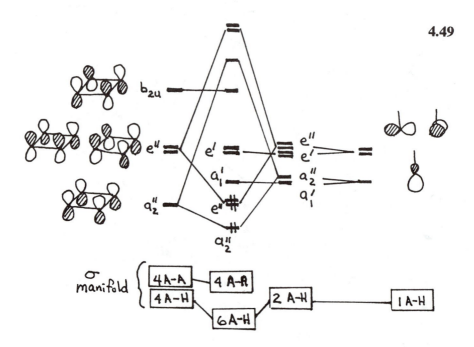

4.49

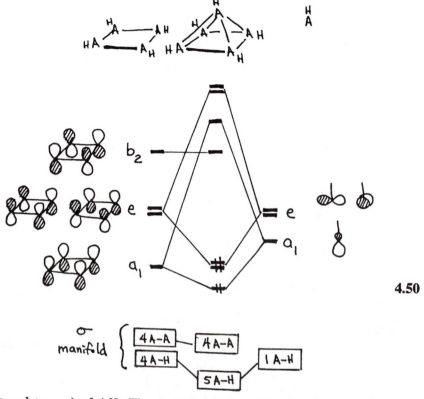

4.50

σ and two π) of AH. These AH orbitals split into pairs symmetric and antisymmetric with respect to the A_4H_4 plane and are labeled respectively a_1' and a_2'', and e' and e'' to reflect this. Only the antisymmetric members of the set ($e'' + a_2''$) may interact with the A_4H_4 π orbitals to give three new skeletal bonding orbitals. When added to the four A-A σ bonding orbitals present initially, this leads to a total of seven skeletal bonding orbitals. For the *nido* octahedron with only e and a_1 orbitals the count is the same (**4.50**).

4.26. (a) In this problem we will use the shorthand notation for the π orbitals of benzene shown in **4.51**. It is easy to construct the symmetry adapted linear

$$\text{[orbital diagram]} \equiv \text{[orbital diagram]}$$

4.51

cominations of these orbitals appropriate for the two benzene units, since these divide into symmetric and antisymmetric pairs. The orbital interaction diagram is then readily generated as in **4.52**. Notice that in this geometry the *HOMO* is pushed up in energy. (b) On bending most orbitals are unperturbed (**4.53**). There are only four that are greatly changed.

Overlap between the a_{2u} π combination and z is lost when θ decreases as in **4.54**.

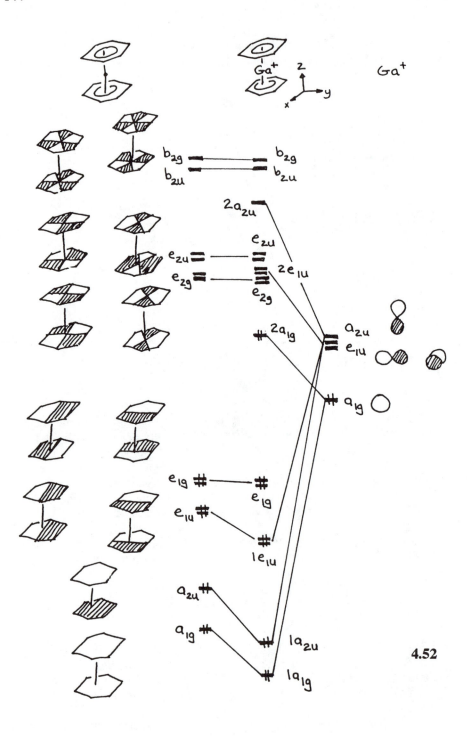

4.52

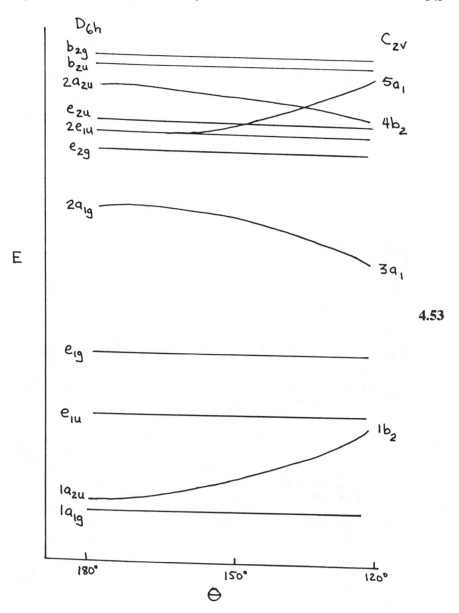

4.53

Therefore, for $1a_{2u}$: $S_{ii} = (-)$ overlap decreases (bonding decreases), $e(1) \propto -S_{ii} = (+)$ and the orbital is destabilized

For $2a_{2u}$: $S_{ii} = (+)$ overlap increases (antibonding decreases), $e(1) \propto -S_{ii} = (-)$ and the orbital is stabilized.

The $2a_{1g}$ orbital lies close to the $2e_u$ set, consequently there can be a second order change for $2a_{1g}$ as shown in **4.55**. $e(2) \propto S_{ij}^2/e_i^0 - e_j^0 = (+)/(-) = (-)$, a stabilization.

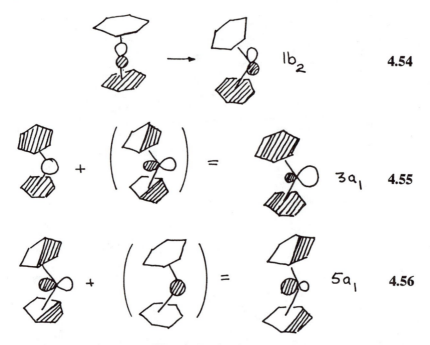

$1b_2$ **4.54**

$3a_1$ **4.55**

$5a_1$ **4.56**

Likewise for $2e_u$. $e^{(2)} \propto S_{ij}^2/e_j^0 - e_i^0 = (+)/(+) = (+)$. *i.e.*, destabilized as shown in **4.56**.

With $3a_1$, the HOMO, decreasing θ is thus stabilizing. We could regard the instability of the 'linear' geometry as arising through a second order Jahn-Teller effect coupling the $2a_{1g}$ and $2e_u$ orbitals as in **4.55** and **4.56**.

(c) The bonding between the benzene ring and Ga^+ comes from overlap of the three filled π orbitals on benzene and the p atomic orbitals on Ga. A CH_3 group is a weak π-donor. Thus, substitution of CH_3 groups results in the π orbitals being raised in energy (**4.57**). A stronger interaction will create shorter Ga-C distances.

4.27. The valence orbitals of a C_{4v} AH_5 species are shown in **4.47**. In SF_5^+ since the F atoms are very electronegative, interactions between the deepest-lying orbitals ($1a_1$ and $1e$) with the orbitals of CH_3^- will be small. The

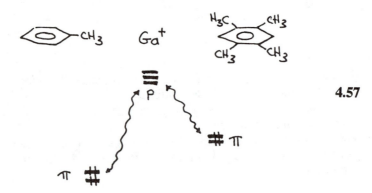

4.57

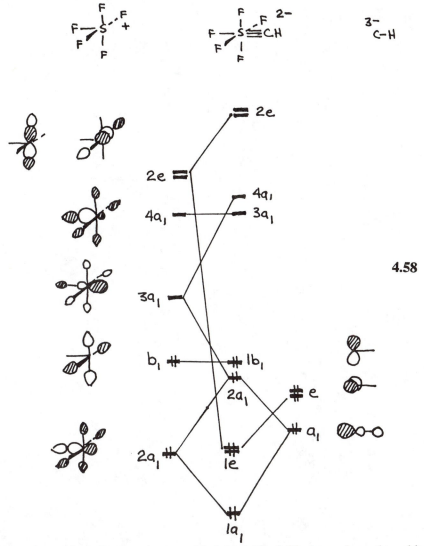

4.58

molecular orbital diagram is assembled in **4.58**. **4.59** shows how the orbital charcter is derived using perturbation theory.

The form of the σ manifold is very similar to that expected for an octahedral AH_6 molecule and thus no immediate bond length difference is expected from this source. However, there is no such symmetry in the π system. Notice that the $e(\pi)$ set of the CH unit mixes with the A-H antibonding $2e$ orbitals of the AH_5 unit. Thus the form of the $1e$ orbital in the molecule is C-H π bonding but is *cis* A-H σ antibonding. As a result we expect that the *trans* S-F bond should be shorter than the *cis* S-F bonds.

4.28. The first task is to assemble a molecular orbital diagram for the square pyramidal B_5H_5 unit. Perhaps the simplest route is from square planar B_4H_4 and BH units. This is an easy exercise once we recognize that overlap with the

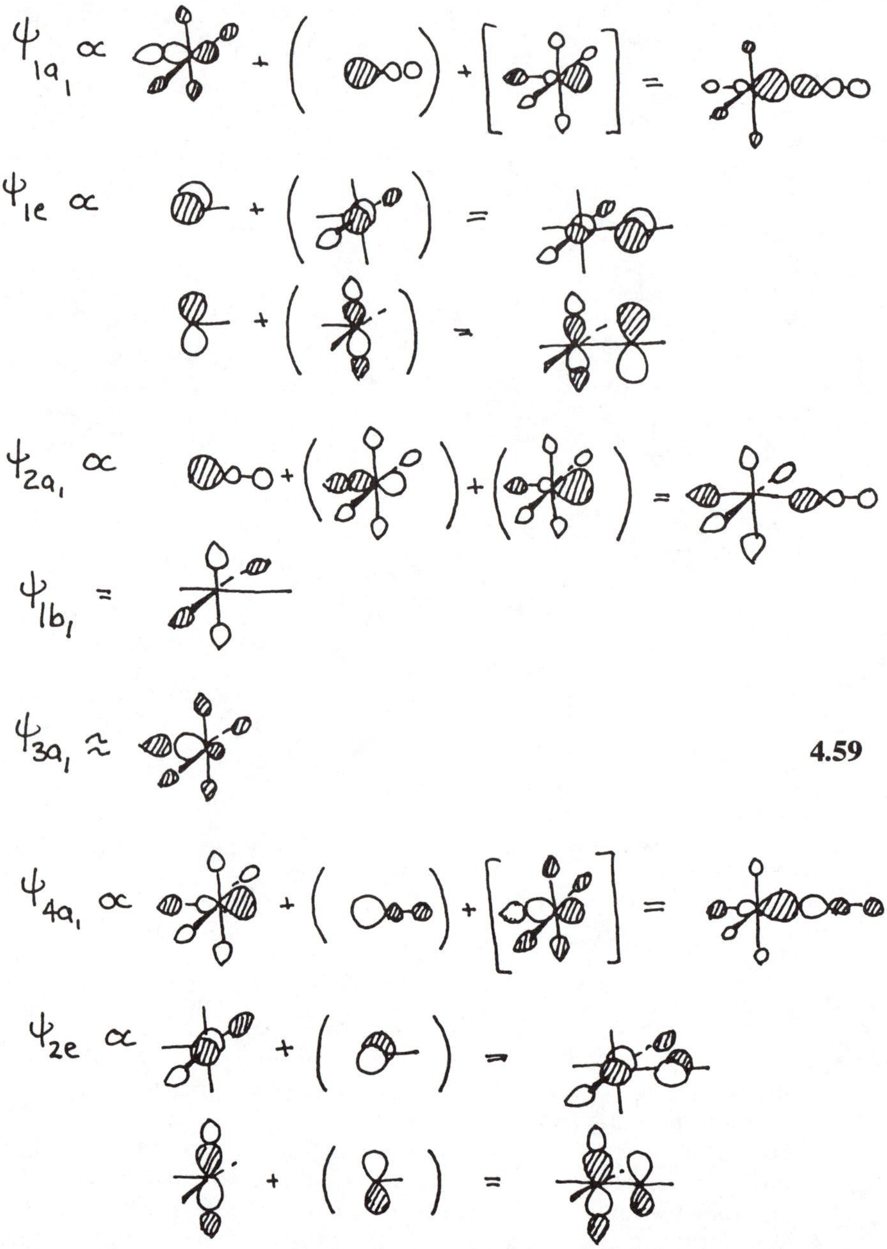

$\psi_{1a_1} \propto$ $+ \left(\text{} \right) + \left[\text{} \right] = $

$\psi_{1e} \propto$ $+ \left(\text{} \right) = $

$+ \left(\text{} \right) = $

$\psi_{2a_1} \propto$ $+ \left(\text{} \right) + \left(\text{} \right) = $

$\psi_{1b_1} = $

$\psi_{3a_1} \approx$

4.59

$\psi_{4a_1} \propto$ $+ \left(\text{} \right) + \left[\text{} \right] = $

$\psi_{2e} \propto$ $+ \left(\text{} \right) = $

$+ \left(\text{} \right) = $

π-type orbitals of the square plane will energetically be the most important since these orbitals point towards those of the *B-H* unit. The assembly process is shown in **4.50**. The eight orbitals associated with the square planar *B-B* and *B-H* bonds lie deeper than those of the π manifold and are so shown. Notice that the level structure for the composite molecule has a good *HOMO-LUMO* gap at 7 skeletal electron pairs, four in *B-B*σ, one in the *a1* and two in the *e* pair of orbitals. This is just the figure suggested by Wade's rules. Now we

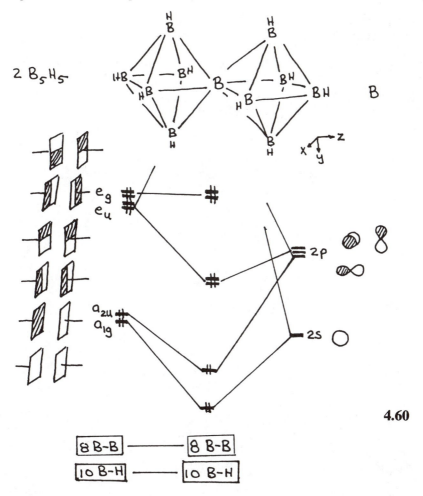

4.60

derive the orbital picture for two vertex-fused octahedra as in **4.60**. With two B_5H_5 units related by a mirror plane all of the orbitals of **4.50** fall into two pairs, one symmetric and one antisymmetric with respect to the mirror plane, as shown at the left hand side of **4.60**. Now the $2s$ and $2p_z$ orbitals of the bridging atom each find a symmetry match with a filled orbital of this set, with the a_{1g} and a_{2u} orbitals respectively, and the $2p_{x,y}$ orbitals link up with the e_u orbital pair. This leads to four B_5H_5-B bonding orbitals. The electrons in the e_g orbital pair, although bonding between apical and basal boron atoms of each unit, remain nonbonding with respect to the fusing atom. The total number of bonding skeletal electron pairs is thus just the same as that for two $B_5H_5{}^{4-}$ units, or two octahedra.

The effect of the size of the fusing atom (E in **4.7**) is an interesting one. The model developed in **4.60** stresses the interaction between the *nido* fragments and the fusing atom. If the latter is small however, as in the boron case, there will also be interactions between the fragments themselves. In fact if all of the close contacts within the two fused octahedra are equidistant, then the inter-octahedral distance is the same as the closest intra-octahedral distances. The result is a more complex picture than that shown in **4.60**.

4.29. The symmetry species of the orbitals of **4.61** are readily generated as in Table 4.9.

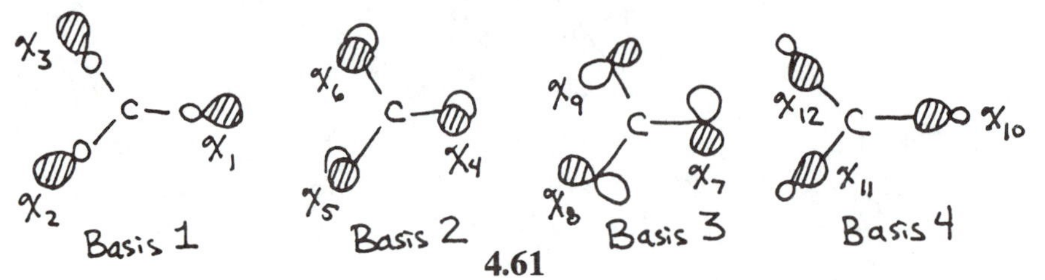

Basis 1   Basis 2   Basis 3   Basis 4

4.61

Table 4.9. Symmetry species of the orbitals of **4.61**.

D_{3h}	E	$2C_3$	$3C_2$	σ_h	$2S_3$	$3\sigma_v$	
Γ_1	3	0	1	3	0	1	$a_1' + e'$
Γ_2	3	0	-1	-3	0	1	$a_2'' + e''$
Γ_3	3	0	-1	3	0	-1	$a_2' + e'$
Γ_4	3	0	1	3	0	1	$a_1' + e'$

The symmetry adapted sets of ligand based orbitals are shown in **4.62**. The relative energy of the a and e sets of each set of orbitals are simply determined by the number of nodes. The energies of the two hybrid sets χ_1 and χ_4 in **4.8** lie deeper in energy than the orbitals composed of pure oxygen $2p$ character. The energy separation of the orbitals derived from the inward pointing hybrids is larger than that for the outward pointing ones on overlap grounds. If

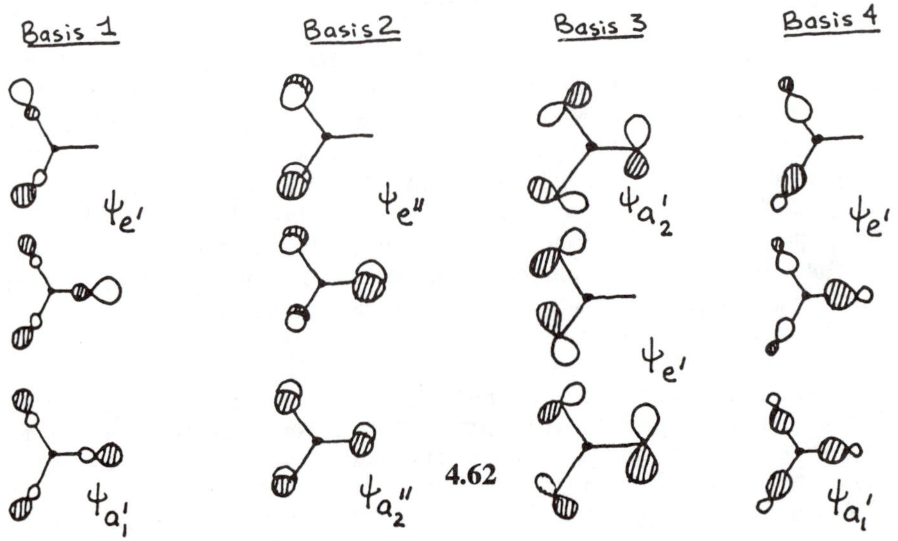

Basis 1   Basis 2   Basis 3   Basis 4

$\psi_{e'}$ $\psi_{e''}$ $\psi_{a_2'}$ $\psi_{e'}$

$\psi_{a_1'}$ $\psi_{a_2''}$ $\psi_{e'}$ $\psi_{a_1'}$

4.62

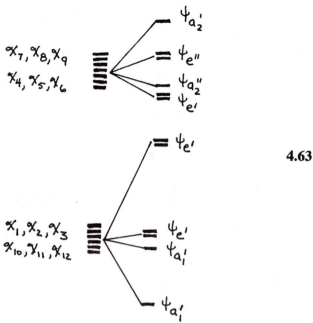

4.63

overlap is switched on between the oxygen atoms then the diagram of **4.63** results.

(b) The orbital diagram is constructed in **4.64**. The strongest interactions are shown for the interaction of the σ orbitals (the inward pointing hybrids, χ_4). Smaller interactions are shown for the orbitals of π type in **4.65** but there is one strong interaction of a_2'' symmetry. There result three σ bonding orbitals ($a_1' + e'$) and one strongly π bonding orbital (a_2''). With a total of 24 electrons a good *HOMO-LUMO* gap is generated.

(c) The π orbitals of e' symmetry compete with the σ orbitals of the same symmetry for central atom character but there are still basically four filled bonding orbitals. Thus the individual $C-O$ bond order is 4/3. This is just

$$\text{(resonance structures)} \quad \textbf{4.66}$$

what is predicted from the resonance structures of **4.66** which must be drawn to satisfy the octet rule. We should not be too carried away by this agreement. The molecular orbital diagram just constructed would be equally applicable to the BF_3 molecule. It has the same geometry and the same number of electrons, but here no resonance structure *need* be drawn to fit the octet rule at fluorine. In other words the bond order is 1. The bond order arguments from molecular orbital theory are still valid though, and predict a bond order of 4/3 if the π electron interactions between the boron and fluorine atoms are important.

4.30. There are $n + 1$ pairs of electrons and n vertices. Thus we may associate

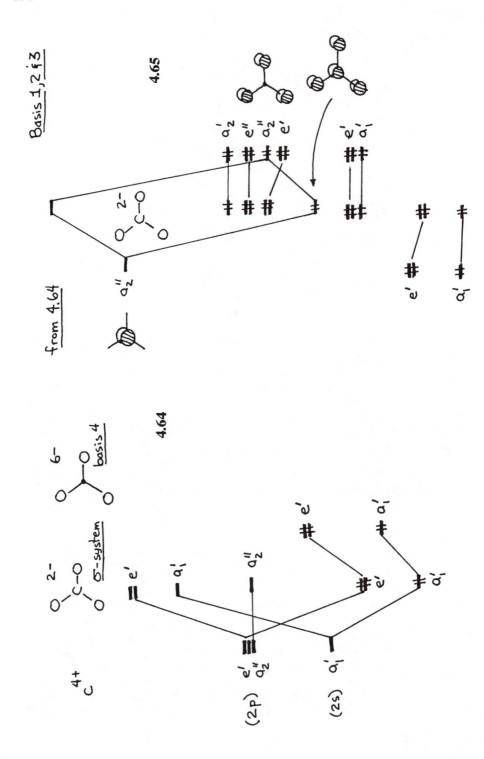

$(n + 1)/n$ electron pairs per vertex. If these electrons are shared equally with all x linkages radiating from a particular x-coordinate vertex, then a measure of the B-B bond order might be $G = (n + 1)/n .((1/x_1) + (1/x_2))$ where the two atoms connected are x_1 and x_2 coordinate. This is plotted in **4.67** for the series of molecules given in **4.9** and shows a reasonable linear correlation for the related *closo* molecules but a different type of plot for the *arachno* and *nido*

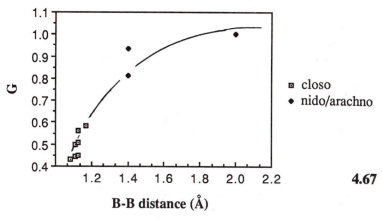

4.67

examples which contain bridging B-H bonds.

4.31. (a) The generation of the molecular orbital correlation diagram for the bending of planar AH_3 is derived in detail in many places. It is shown in **4.34**. The $1e'$ and $2e'$ levels are destabilized and stabilized respectively on bending as a result of the change of overlap between A and H located orbitals on bending. The a_2'' orbital is strongly stabilized on bending as a result of heavy mixing with the $2a_1'$ orbital, which in turn is strongly destablized. (b) NH_3 is a pyramidal molecule in its electronic ground state $(1a_1)^2(1e)^4(2a_1')^2$ since the HOMO ($2a_1$), usually regarded as the energetically most important orbital is stabilized on distortion to the pyramid. We do however know that the barrier to inversion is small ($\sim$5 kcal/mole) which leads to the interesting effects which form the basis for Question 4.11. Promotion of an electron into the $2a_1'/3a_1$ orbital which is strongly destabilized on bending changes the picture. Now the molecule should be planar, which it is. The symmetry species of the possible electronic states is given simply by the direct product $a_2'' \times a_1' = a_2''$, which leads to $^1A_2''$ and $^3A_2''$ states. Promotion of an electron to the next highest molecular orbital ($2e$) should lead to a non-planar geometry, since this orbital has the same energetic tendency on bending as does the HOMO of the ground state of NH_3. The symmetry species of the possible electronic states arise *via* the direct product $a_1 \times e = e$, giving ^1E and ^3E electronic states. Both of these are Jahn-Teller unstable, the symmetry species of the Jahn-Teller active mode being given by the symmetric direct product of $e \times e$. This is evaluated as e in Table 4.10.

     A bending mode of e species is required; motion of this type is shown in **4.68** and takes the C_{3v} molecule to one of C_s symmetry. These results are somewhat hypothetical since the excited states of ammonia are Rydberg states. These are states where the excited electron is localized in an *atomic* orbital of

Table 4.10. The symmetric direct product of *e*.

for e	*E*	*2C₃*	*3σᵥ*	
$\chi(\mathcal{R})$	2	-1	0	
$\chi(\mathcal{R}^2)$	2	-1	2	
$\chi^2(\mathcal{R})$	4	1	0	
$\chi^2_{sym}(\mathcal{R})$	3	0	1	$a_1 + e$

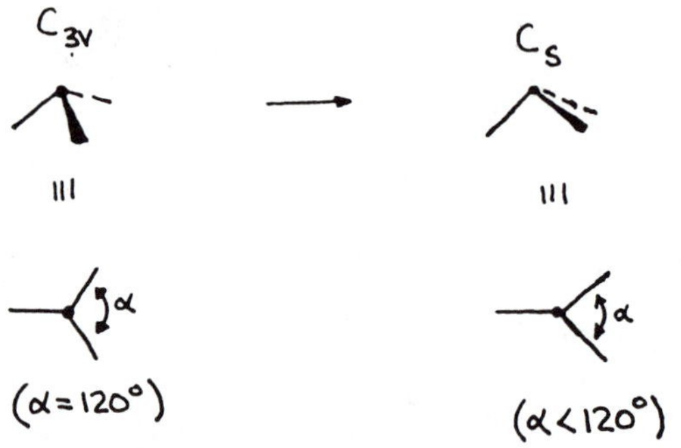

4.68

the molecule. Thus the first excited state of NH_3 is in fact a state with the electron configuration $...(2a_1')^1(3s\ a_1')^1$. Here the second electron is localized in an orbital which has properties similar to those of a *3s* orbital of a one-electron atom, but where the 'nucleus' is the NH_3^+ moiety. This is a system where the Mulliken-Hund, delocalized molecular orbital, model describes some of the electrons $((2a_1')^1$ and deeper), but the localized Heitler-London model describes the highest energy electron. See Question 8.3 for a similar situation in solids.

(c) For the integral $<X/\mu/Y>$ to be nonzero, the product $\Gamma_X \times \Gamma_Y$ must contain Γ_μ. In the D_{3h} point group $\Gamma_\mu \rightarrow e' + a_2''$; in the C_{3v} point group $\Gamma_\mu \rightarrow e + a_1'$. For the planar geometry $\Gamma_X \times \Gamma_A = a_1' \times a_2'' = a_2''$, and $\Gamma_X \times \Gamma_B = a_1' \times e' = e'$. Thus both of the transitions are allowed in this geometry. The situation is the same in the pyramidal geometry. Here on moving from planar to pyramidal, $a_2'' \rightarrow a_1'$ and $e' \rightarrow e$.

4.32. The *VSEPR* approach as used in Question 4.35 is very useful in terms of local geometry determination. However, it is not a viable method to assess the conformational preferences of one part of a molecule relative to another. This

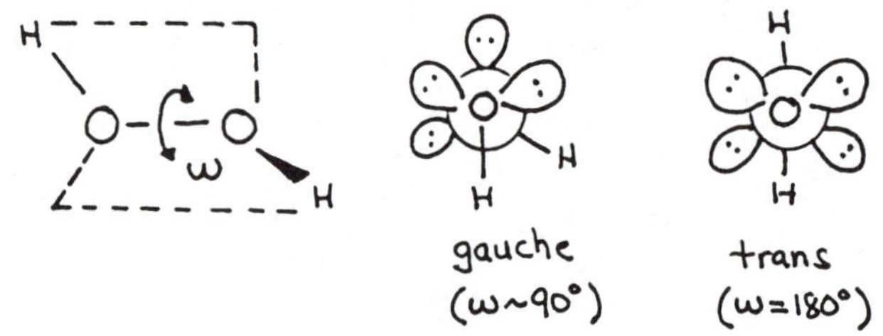

gauche
$(\omega \sim 90°)$

trans
$(\omega = 180°)$

4.69

is illustrated in **4.69** where lone pair-lone pair repulsions are minimized for a torsional angle of 180°. Although this geometry is indeed found for H_2O_2 in some solids containing hydrogen peroxide of crystallization, the lowest energy structure of the isolated molecule is the one with a *gauche* interaction between the lone pairs. Although the α and β values for the different orbitals and their interactions are certainly different, by setting them equal we can capture the essence of the problem. In the *trans* ($\omega = 180°$) conformation, one pair of $p\pi$ orbitals form bonding and antibonding partners, leaving the other $p\pi$ pair to take part in a four orbital *H-O-O-H* delocalized system (**4.70**). For the *gauche* ($\omega = 90°$) conformation there are two perpendicular *H-O-O* systems. Using the results for the Hückel patterns of Chapter 1.11, the two orbital pictures may be written down as in **4.71**. Notice that in the *gauche* form, bonding and non-bonding orbitals are occupied, but in the *trans* form, bonding and antibonding orbitals. Within the framework of simple Hückel theory the observed (*gauche*) form is stabilized over the *trans* form by 0.4β. Obviously the connection between the two halves of the molecule is *via* the $p\pi$-$p\pi$ interaction of the two

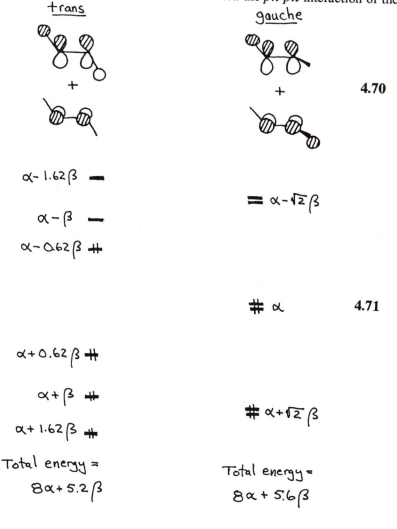

trans

gauche

$+$

$+$ **4.70**

$\alpha - 1.62\,\beta$ ▬

$= \alpha - \sqrt{2}\,\beta$

$\alpha - \beta$ ▬

$\alpha - 0.62\,\beta$ ⧢

⧣ α **4.71**

$\alpha + 0.62\,\beta$ ⧢

$\alpha + \beta$ ⧢

⧣ $\alpha + \sqrt{2}\,\beta$

$\alpha + 1.62\,\beta$ ⧢

Total energy =
8α + 5.2β

Total energy =
8α + 5.6β

oxygen atoms. As this becomes weaker (*i.e.*, if the oxygen is replaced by a heavier main group atom) then the *gauche* effect ought to be less pronounced. It is.

4.33. Each *F* contributes *1* electron. Along with 7 from *I* there are a total of *14e⁻* or *7e⁻* pairs around the iodine atom. Therefore *VSEPR* predicts a pentagonal bipyramid, D_{5h}, or monocapped octahedron, C_{3v}. The orbital diagram for the pentagonal bipyramid is constructed in **4.72**. Notice that four bonding and four antibonding levels are produced. Left behind are three nonbonding levels of a_1' and e_1' symmetry. The axial *I-F* bonds are clearly three center-two electron ones, but the equatorial bonds perhaps best described as 'six center-two electron' ones. On a *p* orbital only model we would say that the axial bond order is 1/2 and the equatorial bond order 2/5. The axial distances should then be shorter than the equatorial ones as found

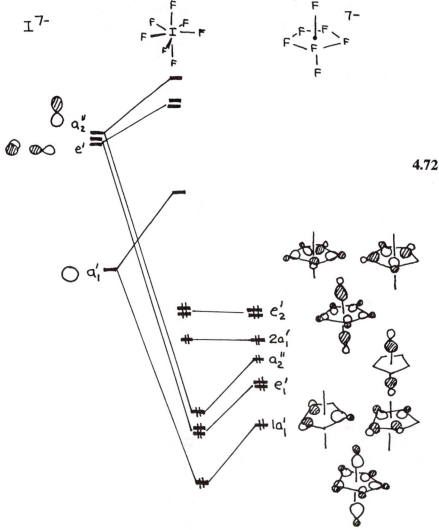

4.72

experimentally.

The orbitals of a capped octahedron can easily be constructed starting with those of the octahedron (**4.73**). The splitting pattern turns out to be very similar.

4.34. (a) Consider first the triangular Te_3^{2+} unit. We expect that there will be three 2 center-2 electron bonds that describe the σ framework, three outward-pointing lone pair orbitals and three π -type orbitals of $a_2'' + e''$ symmetry. The latter two sets are shown in **4.74**. The lone pair orbitals will be at a lower

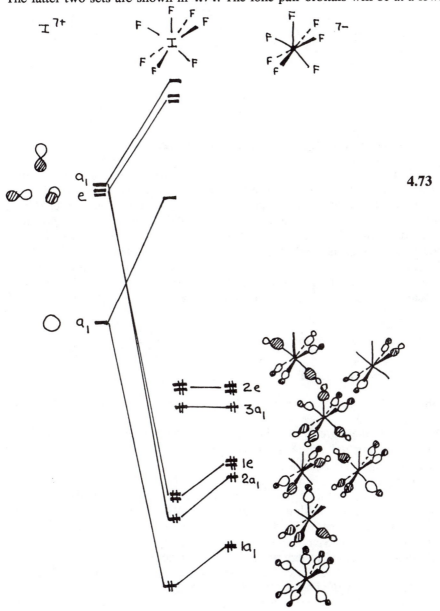

4.73

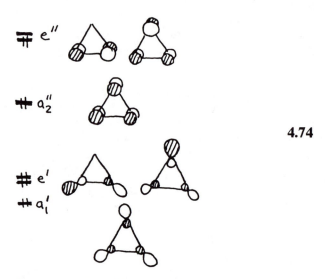

4.74

energy since they are sp hybrids; recall the situation for a bent, C_{2v} AH_2 system.

In the assembly of the orbital diagram for the six-atom unit, the overlap between the in-plane fragment orbitals on one triangle with those of the other will be small, but that between the two sets of a_2'' and e'' orbitals is of σ type and the interaction will be large (**4.75**). Therefore, focusing on these interactions the diagram of **4.76** results. Notice that the antibonding a_2'' molecular orbital is doubly occupied in Te_6^{4+}. This should lead to an elongation of the Te-Te distance between the two triangles when compared to those within each triangle. A valence bond approach would say that the three resonance structures of **4.77** contribute, where each solid line represents a 2

4.75

center-2 electron bond. In molecular orbital terms the inter-triangle bond order is only 2/3.

(b) In C_6H_6 there are C_3H_3 triangles. Now the electron counting is such that the antibonding a_2'' orbital is empty. The bond order of the inter- and intra-triangle linkages are equal and $a = b$.

4.35. (a) For I_2Cl_6 the terminal Cl groups use 1 electron to pair up with one electron from I. The bridging Cl atoms formally contribute 1 electron to one I-Cl linkage and a pair of electrons to the other from a Cl lone pair, effectively a $3e^-$ donor (**4.78**). Since each I contributes 7 electrons and the surrounding chlorines contribute 5, there are therefore 12 electrons around each I; *i.e.*, six electron pairs. Therefore the local geometry around each I should be octahedral as in XeF_4 and structure **a** favored.

(b) For $Hg_2Br_6^{2-}$ there are again 5 electrons from the three Br atoms

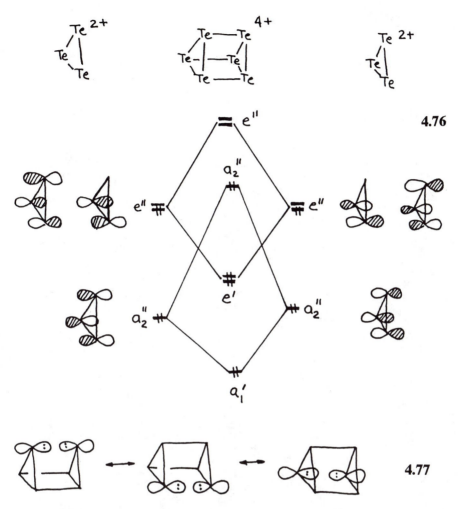

4.76

4.77

surrounding each mercury. Hg^{1-} has the configuration $5d^{10}6s^26p^1$, the ten d electrons are considered "core-like" (and complete a spherical shell), thus there are $3+5 = 8$ electrons around each Hg atom. These four electron pairs should give rise to a tetrahedral environment around the metals (*i.e.*, structure **b**). This problem shows that we don't need molecular orbital arguments to come up with geometry predictions in many main group systems. Electron counting plus the knowledge of some rules of thumb (*VSEPR*) are often sufficient (but see Question 4.32.)

4.78

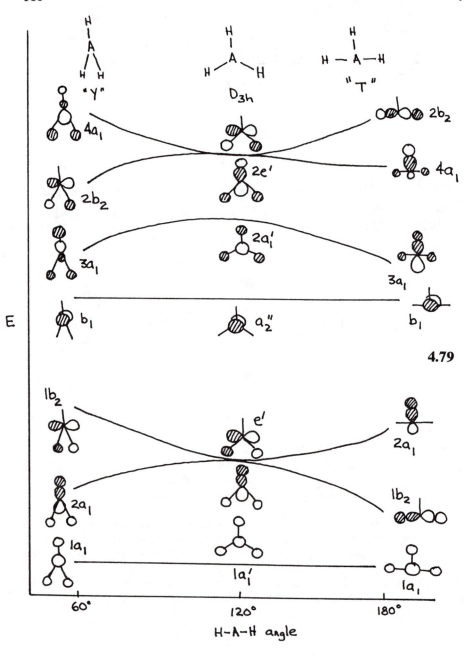

4.36. (a) The orbital correlation diagram is readily constructed as in **4.79**. (b) There are three electron counts for which the *HOMO* has a negative slope. With 4 electrons either C_{2v} geometry should be stable, *i.e.*, $(1a_1)^2(2a_1)^2$ or $(1a_1)^2(1b_2)^2$. With 10 electrons note that there will be a second order energy stabilization between $2a_1'$ and $2e'$. There are large coefficients in these antibonding orbitals, thus either geometry should be stabilized. With 12

e'

4.80

(C_{2v})

(C_{2v})

electrons only one C_{2v} geometry is stabilized. At the T-shape geometry the first order stabilization of $4a_1$ is offset by the second order destabilization with $3a_1$. (c) LiH_3 at a D_{3h} geometry has the electronic configuration $(1a_1)^2(1e')^2$. Therefore, a first order Jahn-Teller distortion is predicted. To evaluate the symmetry species of the Jahn-Teller active distortion coordinate we need the symmetric direct product of e'. This is shown in Table 4.10 for the e species in the C_{3v} point group in Question 4.31. The result is a similar one here and a distortion of e' symmetry results shown in **4.80**. This is the same conclusion as that obtained by consideration of the orbital correlation diagram (as, of course, it should be).
(d) From the results above, a least motion path is not preferred. Rather a path through the Y C_{2v} geometry as the transition state will occur. The potential energy surface will be a "Mexican-hat" surface like that on p. 101 of Reference 1. A, B & C are the three T geometries, D is the Y transition state and E is the D_{3h} mountain top.

4.37. The σ-only molecular orbital diagram for the octahedron is readily derived as in **4.15**. The electronic configuration for $n = 1$ with six valence electron pairs is $(1a_{1g})^2(1t_{1u})^6(e_g)^4$. This configuration fills all A-X bonding orbitals and places two pairs in the e_g nonbonding orbitals. For $n = 3$ the electron configuration is $(1a_{1g})^2(1t_{1u})^6(e_g)^4(2a_{1g})^2$. These extra two electrons occupy an A-X antibonding orbital and the bond length is expected to increase compared to the $n = 1$ case.

4.38. This process is shown in **4.81**. We have used localized orbitals on each BH_2 unit but the same picture will arise using the delocalized ones as a basis. The orbitals of the B_2H_4 molecule itself are readily ordered in energy. After this step there is certainly a σ bond between the two boron atoms with two electrons in the a_g orbital. The *LUMO*, the π-bonding orbital in ethylene with two more electrons is unoccupied here. We could envisage a higher energy state of B_2H_4 with two electrons in this π orbital and no electrons in the σ-bonding orbital. Both arrangements would involve a formal B-B bond. In the second step both of these B-B bonding orbitals mix strongly with the relevant combinations of the bridging H_2 unit. A B-B π-bonding orbital $(1b_{1u})$ is created at low energy as a result of interaction with the out-of-phase combination of the bridging H_2 unit and is now doubly occupied as is the B-B σ-bonding orbital $(1a_g)$. These are now orbitals strongly bonding between both

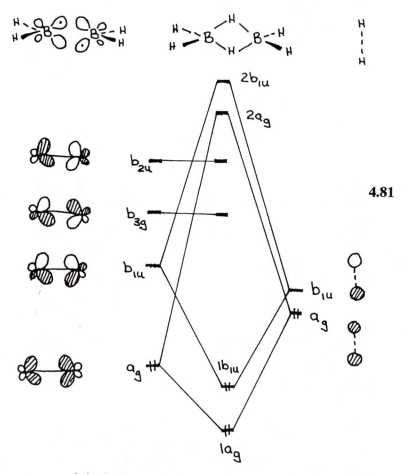

4.81

boron atoms and the hydrogen atoms. They are still boron-boron bonding but now the boron $2p$ character has been diluted by mixing with hydrogen $1s$ and thus the contribution to the B-B overlap population from this orbital will be reduced. But there is still weak σ- and π-type bonding interactions between the boron atoms.

4.39. The point group of the molecule is D_{3h} and **4.82** shows the assembly of the π molecular orbital diagram using the two sets of a_1' and e' orbitals. Since N is more electronegative than B, the bonding orbitals, and thus the electron density, is preferentially located on nitrogen. (Also see question 6.19.)

4.40. Molecular orbital diagrams for the XeF_n series are given in **4.15**. Notice that the *HOMO* in each case is the out-of-phase combination of n ligand σ hybrids with the Xe $6s$ orbital. The *LUMO* lies at the same energy in all three molecules, being just an antibonding combination of a Xe $6p$ orbital with the relevant pair of fluorine σ hybrids (**4.83**). The *HOMO* is increasingly destabilized as n increases and as a result the *HOMO-LUMO* gap decreases. As n increases therefore these XeF_n molecules become increasingly susceptible to

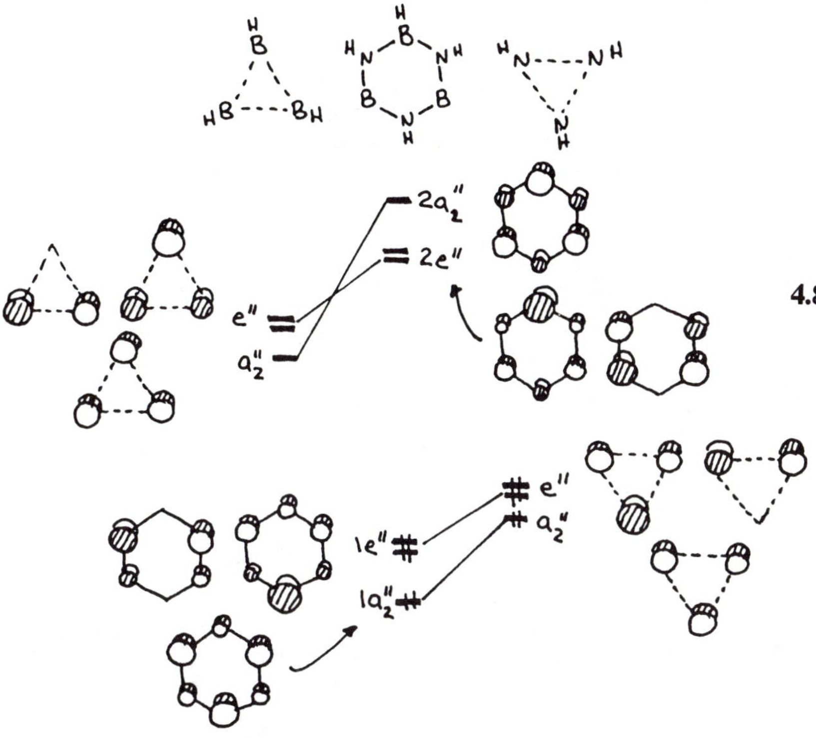

4.82

a second order Jahn-Teller distortion on energy grounds. For XeF_6 the symmetry species of the distortion coordinate is $a_{1g} \times t_{1u} = t_{1u}$, for XeF_4 $a_{1g} \times e_u = e_u$, and for XeF_2 $\sigma_g^+ \times \sigma_u^+ = \sigma_u^+$. Distortions of the octahedron and square plane of these symmetry species change the angular geometry as shown in **4.84**, but not in XeF_2. In practice a distortion is only found experimentally for XeF_6, a result in accord with the difference in *HOMO-LUMO* gap between the two molecules.

Homo   Lumo

4.83 (for XeF_2)

4.84

4.41. First let us see what simple electron counting arguments would predict for these geometries. $ArFCl$ is a twenty-two electron molecule and should thus be linear, which it is. Similarly $ArCO_2$ is isoelectronic with the the carbonate ion, and is planar, and $ArBF_3$ is trigonal pyramidal, being isoelectronic with $ClCF_3$. However, should the Ar atom be coordinated to the F or Cl end of ClF? As shown in Question 4.10 the site preferences for the most electronegative atoms in a molecule depend upon electron count. We can use a simple argument to decide on the geometries here. An orbital interaction diagram linking the *HOMO* of the Ar atom and the *LUMO* of either ClF, CO_2 or BF_3

can easily be constructed. The interaction between the two frontier orbitals will depend upon the site of attachment of the *Ar* atom. It will be maximized by ensuring that the *Ar* atom orbitals overlap with the orbitals of the atom which make the largest contribution to the *LUMO*. The situation is very similar to that involving $Cr(CO)_5$ in Question 3.14. If we recall that bonding orbitals are largely located on the more electronegative atom and antibonding orbitals on the least electronegative atom, then such an argument suggests that for these electron counts the most stable geometry will arise *via* attachment of the electronegative *Ar* atom to the least electronegative atom of the parent, namely *Cl*, *C* and *B* respectively. Indeed this is the case experimentally. In fact *Ar-Cl-F* is isoelectronic with a whole series of trihalide ions such as *Br-Br-Cl⁻* where the more electronegative atom is on the end. The interaction diagrams for $ArBF_3$ and $ArCO_2$ are shown in **4.85** and **4.86** respectively.

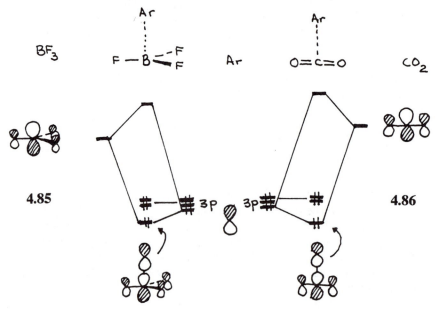

4.42. This is a variant of Question 4.20. On going from CH_3^- to NH_3 to OH_3^+ the central atom becomes more electronegative. Thus the energy of the HOMO drops in this order and the mixing of $2a_1'$ into a_2'' is energetically less stabilizing along this series. As a result the equilibrium *H-A-H* angle becomes larger and the inversion barrier drops. On the other hand, as *A* becomes more electronegative the hydrogen coefficients in $2a_1'$ increase and this will lead to an increase in the overlap between $2a_1'$ and a_2'' as the electronegativity of *A* increases. The energy gap dominates in this case over overlap and this competition may be why the effect is not a particularly large one.

4.43. (a) **4.87** shows the construction of the diagram for CO_2. For the σ manifold we consider the $2s$ and $2p_z$ orbitals on carbon and the s/p hybrid orbitals on each oxygen. They lie energetically between the free atom $2s$ and $2p$ orbitals. The outward pointing hybrids (which transform as $\sigma_g^+ + \sigma_u^+$) have negligible overlap with the central atom orbitals and remain as lone pairs.

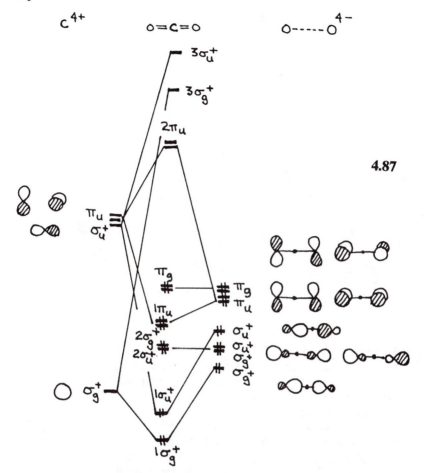

The symmetric and antisymmetric combinations ($\sigma_g{}^+ + \sigma_u{}^+$) of the pair of inward pointing hybrids are of the correct symmetry to interact with the $2s$ and $2p_z$ orbitals on carbon. Only one oxygen π combination finds a symmetry match with a central atom orbital, leading to a pair of bonding and antibonding orbitals.

(b) The *HOMO* (π_g) and *LUMO* ($2\pi_u$) of CO_2 are not of the correct symmetry to mix on bending. We may test for this using second order Jahn-Teller ideas, by seeing if their direct product contains π_u, the symmetry species of the bending mode. In fact $\pi_u \times \pi_u = \sigma_u{}^+ + \sigma_u{}^- + \delta_u$. However with two more electrons the *HOMO* and *LUMO* symmetries are π_u and $\sigma_g{}^+$. Now their direct product does contain π_u (in fact $\sigma_g{}^+ \times \pi_u = \pi_u$) and the molecule should bend. The orbital mixing and energetic stabilization is shown in **4.88**. With two electrons in the *HOMO*, SO_2 and O_3 should have a smaller bond angle than NO_2 with only one electron. Experimentally the bond angles for the gaseous molecules are found to be 119.5°, 116.8° and 134.1° respectively.

(c) This is drawn out in **4.89** for the case of NO_2. The picture is very similar to that derived in (b). The smaller oxygen character for the $CO_2{}^-$ case reflects the greater electronegativity difference between C and O than between N and

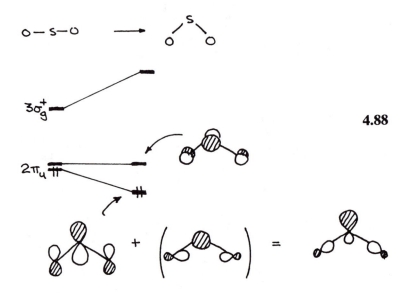

4.88

O. (See also question 4.2.)

(d) The diagram is shown in **4.90**. The overlap between the two orbitals will depend upon the central atom orbital contribution. If the terminal atoms are much more electronegative than the central atom then the electron density will be largely localized on the central atom and the overlap between the two units will be large. This is expected to be the case for BF_2. If the electronegativity of the atoms are similar (as in NO_2) then the overlap will be considerably smaller since the central atom coefficient is much reduced. In this case the interaction will be much weaker. Thus the $B\text{-}B$ bond is stronger than the $N\text{-}N$ bond.

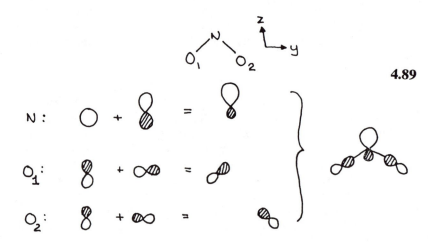

4.89

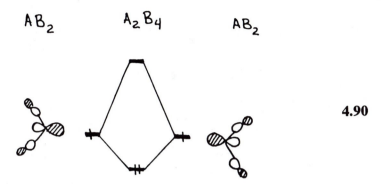

4.90

Chapter V

Organometallic Chemistry

5.1. By comparing molecular orbital diagrams for square-planar and octahedral ML_4 and ML_6 complexes show the origin of the octahedral eighteen-, and square-planar sixteen-electron rules. (Use a σ-only orbital model.) Hence provide a molecular orbital explanation for the observation that whereas $V(CO)_6$ and $Cr(CO)_6$ are stable molecules, $Fe(CO)_6$ does not exist.

5.2. Devise organometallic compounds using the isolobal analogy which correspond to the following. (a) bicyclobutane (b) cyclopropenium cation (c) P_4 (d) methyl cyanide (e) allene.

5.3. Explain why the CO stretching vibrational frequencies increase in the order $Cr(CO)_6 < Fe(CO)_5 < Ni(CO)_4$.

5.4. Determine the symmetry properties of the π orbitals of the acetylene units coordinated to tungsten in the molecule $(C_2H_2)_3W(CO)$. Hence provide an explanation of its apparent twenty electron count.

5.5. Using the frontier orbitals of a $Cr(CO)_5$ unit and the orbitals of the H_2 molecule, construct a molecular orbital diagram for the species $Cr(CO)_5(H_2)$ containing a sideways coordinated dihydrogen. You will have to label the symmetry species of the orbitals in a fashion appropriate to the C_{2v} point group. Does your orbital picture predict a good $HOMO\text{-}LUMO$ gap for this molecule? It is found that such non-classical or dihydrogen complexes $(ML_n(H_2)$ species) are often found when some of the ligands attached to the metal are good π-acceptors, or when the molecule has a positive charge. With poorer acceptors or neutral molecules the classical dihydride is found. Using your molecular model show how the nature of these ligands controls the extent of the mixing of the H_2 σ^* orbital into the metal located orbitals. Hence provide an orbital explanation of this result.

5.6. Explain the following observations: (a) The width of the lowest-lying electronic transition in the $V(CO)_6$ molecule is considerably larger than that for $Cr(CO)_6$. (b) Low-spin d^8 molecules are never found in octahedral geometries. (c) $Cr(CO)_6$ comes in a dark bottle from the supplier.

5.7. Carbon monoxide binds to transition metals via the carbon end. Devise a model to rationalize the non-observation of the (oxygen bound) isocarbonyl.

What does your scheme have to say about the strength of coordinated N_2?

5.8. **5.1** shows a molecule made a few years ago by Huttner and coworkers. It may be regarded as a diatomic As_2 molecule with three $Cr(CO)_5$ units coordinated around the axis. Use molecular orbital arguments to decide on the bond order between the As atoms in the following way.

5.1

5.2

a

b

(a) Construct a molecular orbital diagram for As_2, labeling the orbitals with the symmetry species appropriate for the point group of the As_2Cr_3 fragment.

(b) Construct symmetry adapted linear combinations of orbitals for the three $Cr(CO)_5$ units coordinated around the axis. Only one orbital, a hybrid of metal s, p and d, which points out of the bottom of the square pyramid is important here. (It is shown in Chapter 1, and is empty in the isolated fragment). So there are only three $Cr(CO)_5$ orbitals to consider.

(c) Assemble the complete diagram for the molecule, add the electrons and work out the bond order.

5.9. The geometries of two ethylene complexes are shown in **5.2**. Determine the electronic preferences for the olefin orientation (in-plane or out-of-plane) by considering the interaction of the C_2H_4 π^* level with the orbitals of the the σ-only model. Are your predictions in accord with experiment? If not, why not?

5.10. Using the frontier orbitals of a square pyramidal ML_5 unit and the orbitals of the triangular H_3 molecule, construct a molecular orbital diagram for the (so far hypothetical) species $M(CO)_5(H_3)$ containing an H_3 molecule coordinated in the sixth site. For which electron count, and thus for which transition metal atom M, might such a species be stable?

5.11. Molecules of the type $(PR_3)_2Pt\text{-}Pt(PR_3)_2$ often have very short $Pt\text{-}Pt$ distances. This is at first sight difficult to understand since each $(PR_3)_2Pt$ unit has a closed shell d^{10} configuration. (By analogy with two closed shell He atoms, the potential between the two fragments might be expected to be repulsive.) Construct a molecular orbital diagram for the dimer using the frontier orbitals of the fragments. Include in your diagram mixing between primarily metal located d and s orbitals of the same symmetry in a similar way to the s/p mixing for first row diatomic molecules. Hence generate a model for a non-zero bond order in this molecule. Compare your result with that which would come directly from use of the isolobal analogy.

5.12. The molecule $(AuPR_3)_6C^{2+}$ is known. It consists of an octahedron of gold atoms with close Au-Au distances. At the centroid of the octahedron is located a carbon atom. The Au-PR_3 unit electronically has great similarities to the simplest of all species, namely the hydrogen atom. Mimick the molecular orbital diagram for the $(AuPR_3)_6$ octahedron by constructing that for the H_6 octahedron. Draw out the form of the molecular orbitals and their approximate relative energies. Insert the orbitals of the carbon atom, add the relevant number of electrons and show that this structure is electronically satisfactory for this molecule.

*5.13. A highly unusual $Os(NR)_3$ molecule was recently synthesized. Its X-ray structure is shown in **5.3**. Idealize the structure by considering $Os(NH)_3$ at a D_{3h} geometry and using all of the lone pairs on the NH units along with the $5d$,

5.3

5.5

$6s$ and $6p$ atomic orbitals on Os construct a molecular orbital diagram and draw out the resulting shapes of the orbitals. The authors who reported the compound regard it as a 20-electron compound. How are they counting electrons? A planar ML_3 compound should contain only 16 electrons. Show why there seem to be four extra electrons.

5.14. Draw out the all-organic (containing just C and H) analogs of the molecules of **5.4** using the isolobal analogy.

5.15. Methyl lithium is a tetramer with the geometry shown in **5.5**. Construct a molecular orbital diagram for the tetramer from the orbitals of Li_4 and Me_4 units. Describe the bonding situation which results. (Hint: Take account of the electronegativity difference between carbon and lithium in constructing the diagram.)

5.16. Use Wade's rules to predict the approximate shape for the following molecules. (a) $Ni_5(CO)_{12}{}^{2+}$, (b) B_5H_9, (c) $H_2Ru_6(CO)_{18}$, (d) $Fe_5(CO)_{15}C$, e) $(CpCo)_2C_2B_6H_8$, f) B_9H_{15}, g) $Co_8C(CO)_{18}{}^{2-}$, h) $(PhC{\equiv}CPh)Ru_4(CO)_{12}$.

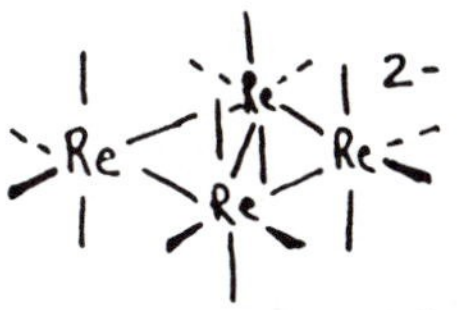

a) $(Re_4(CO)_{16}^{2-})$

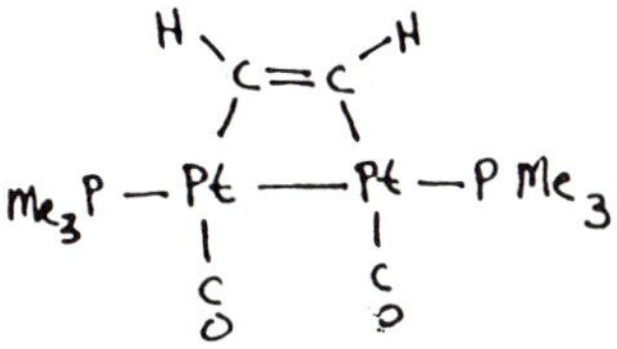

b)

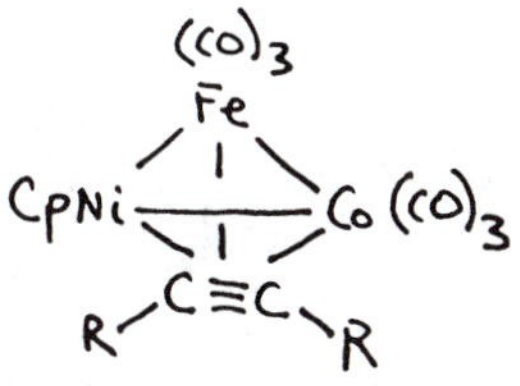

e)

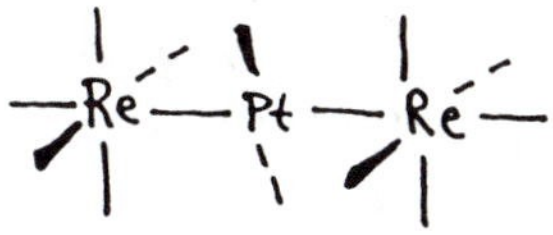

d) $(Re_2Pt(CO)_{12})$

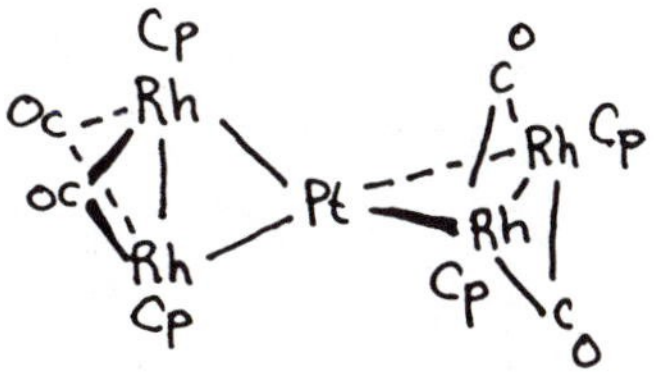

e)

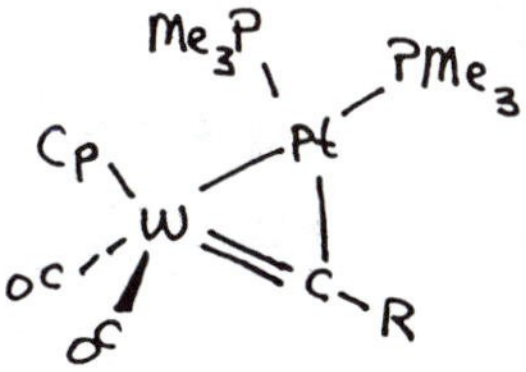

f)

5.4

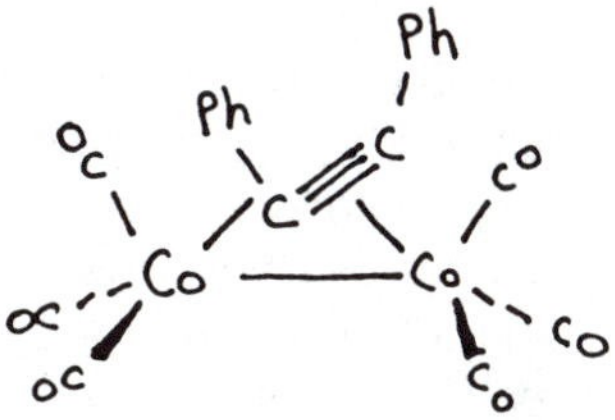

g)

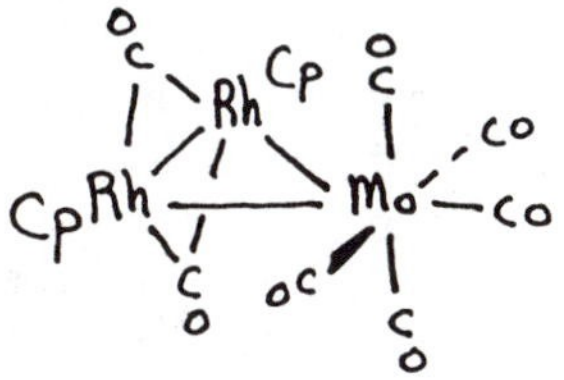

h)

There are two possibilities for (d) and (g).

5.17. (a) Determine the important valence orbitals for an $(\eta^6\text{-}benzene)M$ complex. (b) Interact the valence orbitals of the $(\eta^6\text{-}benzene)M$ fragment with three ligands to produce a C_{3v} $(\eta^6\text{-}benzene)ML_3$ molecule. (c) Show that the electronic structure of this molecule is analogous to that in ferrocene or octahedral ML_6.

5.6

5.18. The molecule in **5.6** may be made in both diamagnetic and paramagnetic forms simply by changing the π-donor nature of the sulfide containing ligand. Assemble a molecular orbital diagram for the species from $NiCp^+$ plus $RSCH_2CH_2SR$ and show how this comes about.

5.19. **5.7** shows the variation in the stability constant for a series of nickel olefin complexes as a function of the energy of the *LUMO* of the olefin. What does this plot tell us about how the olefin is bonded to the metal?

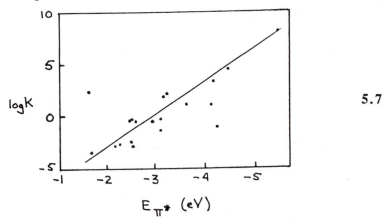

5.7

5.20. Shown in **5.8** is the structure for a $M(NH_2)_4$ molecule. All *M-N* and *N-H* distances along with *N-M-N* angles are identical. All M-N-H angles are 120°. (a) Using the *p* atomic orbitals on nitrogen, labeled $\chi_1-\chi_4$ in the figure, form normalized symmetry adapted combinations and draw them out. (b) Consider *M* to possess only *s* and *p* valence atomic orbitals. Construct an orbital interaction between these atomic orbitals and the symmetry adapted linear combinations from (a). As a guide, consider that H_{mm} for $Ms = -8.0eV$, $Mp = -6.0eV$ and $Np = -13.0eV$. (c) Draw out the resultant molecular orbitals from (b).

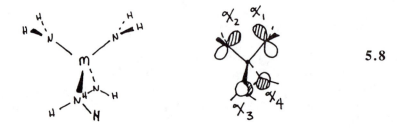

5.8

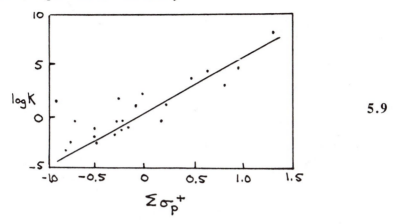

5.9

5.21. **5.9** shows the observed correlation between the stability constant for the formation of a series of olefin complexes of *Ni* with the sum of the Hammett σ constants of the substituents attached to the olefin. (This parameter measures the electron withdrawing propensity of the ligands.) For an analogous series of complexes of Ag^+ the plot has the opposite slope. What do these data tell us about how the olefin is bonded to the metal in each case?

5.22. Produce a molecular orbital argument to decide whether the coordinated olefin in trigonal bipyramidal $Fe(CO)_4(C_2H_4)$ will lie parallel to, or perpendicular to the 'trigonal' plane.

5.23. (a) Derive a molecular orbital diagram for ferrocene in the staggered geometry by using the π orbitals only on the ring atoms. (b) Show that an 18 electron count arises for this molecule. (c) Predict the spin multiplicity for the 20-electron molecule nickelocene. (d) The properties of the species containing manganese depend upon the nature of the ligand atoms attached to the ring. Rationalize the observations that for Cp_2Mn a hextuplet is found with a long *M-C* distance, but $(Me_5Cp)_2Mn$ is a doublet with a shorter *Mn-C* distance.

*5.24. The structure of $Pd[Sn(NR_2)_2]_3, R = Me_3Si$, is shown in **5.10a**. If one discounts the presence of the *R* groups, then the geometry is D_{3h}. (a) Making the approximation that the NR_2 groups behave like hydrogens, draw an interaction diagram for the $Pd(SnH_2)_3$ molecule considering only σ interactions between *Pd* and *Sn* for a *Pd* atom interacting with the $(SnH_2)_3$ unit. (Hint: Consider SnH_2 to have about the same electronegativity as *Pd*.) In your diagram show which orbitals are filled. (b) Draw out explicitly the metal *d* centered orbitals. (c) Using the orbitals from (b) and their relative energies from (a), now allow for π interactions between *Pd* and *Sn*. (d) An alternative structure which also has D_{3h} symmetry is shown in **b**. Discounting the increased steric interactions between the *R* groups, show in detail why structure **b** is likely to be less stable than **a**.

5.25. Use only the most important valence orbitals in $(CH_2)_2PtCl_2$ to explain why conformation **a** is more stable than conformation **b** in **5.11**.

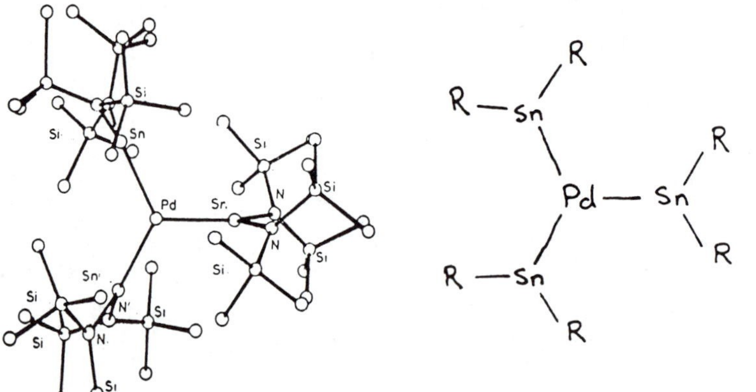

a   **5.10**   b

5.26. Construct a Walsh diagram connecting linear and bent ML_2 molecules and hence show how one is stable for 16 and the other for 14 electrons.

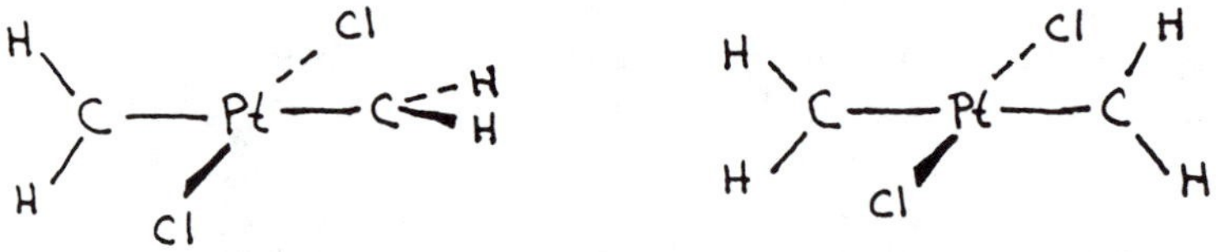

a   **5.11**   b

5.27. The isolobal analogy is often useful in correlating the structures of apparently different species, but sometimes the structures of a pair of molecules related in this way are markedly different. $AuPR_3$ and H are isolobal but the structure of $N(AuPR_3)_5^{2+}$ (**5.12**) is quite different from that of CH_5^+ (Question 6.10.). Construct a molecular orbital diagram for the gold-containing species and comment on the factors which might stabilize one geometry over another.

5.28. The CR_2 ligand in molecules of the type Cp_2WLCR_2 may be oriented in the two possible ways shown in **5.13**. Steric repulsion between the rings and the CR_2 group obviously favor **b**, but which of the two orientations is prefered

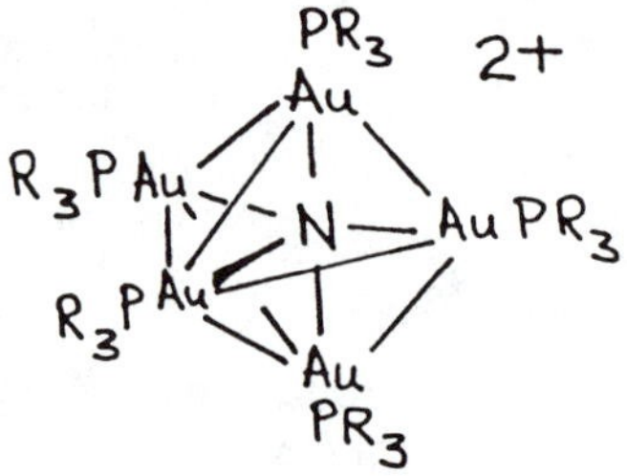

5.12

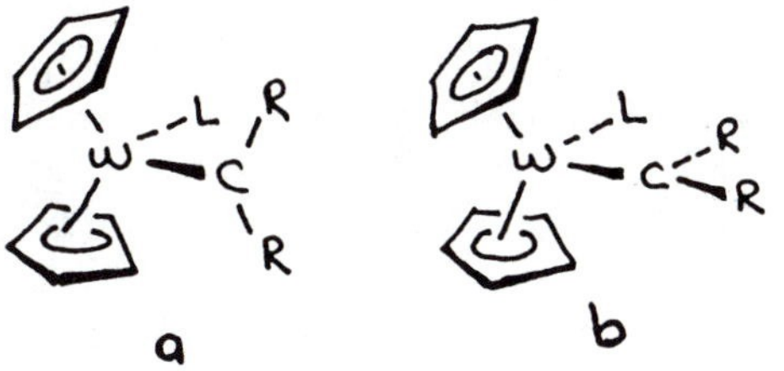

5.13

electronically for the d^2 system?

*5.29. **5.14** shows the chemistry of the Dötz reaction. A critical intermediate in this process has been suggested to be the metallacyclobutene shown in **5.15**. (a) By simple electron counting show that this species is unsaturated. (b) Construct a molecular orbital diagram for a model compound using the

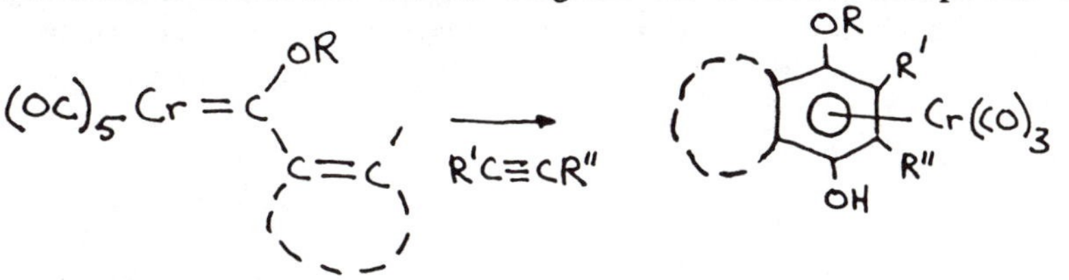

5.14

frontier orbitals of the $Cr(CO)_4$ and $CH_2\text{-}CH\text{=}CH$ fragments. Hence show that this species is short of two electrons for stability. (c) Show that the species shown in **5.16** is a more likely structure for the crucial Dötz intermediate.

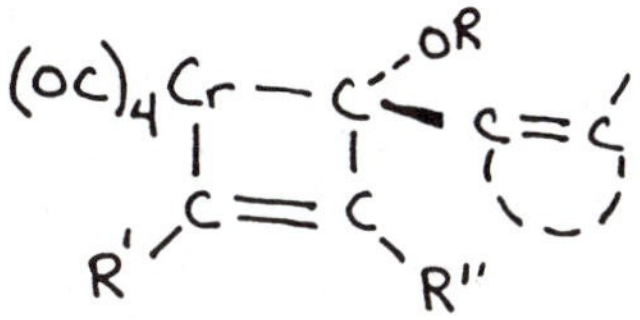

5.15

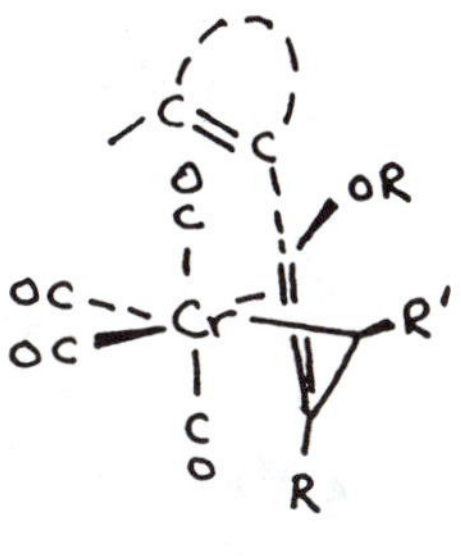

5.16

5.30. Determine a simple reason why the *Rh-Rh* bond length changes for the three geometrical isomers of **5.17**.

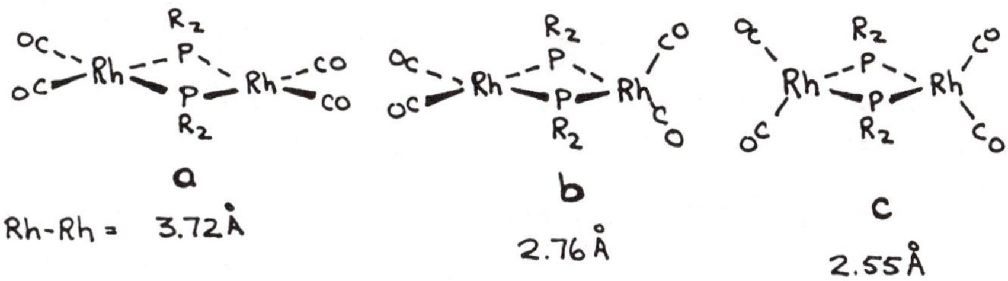

Rh-Rh =   3.72Å

2.76 Å

2.55Å

5.17

5.31. Construct a molecular orbital diagram for the interaction of a $Fe(CO)_3$ fragment with an η^4 butadiene molecule. By looking at the form and electron occupation of the molecular orbitals which you generate, provide a rationale for the experimental observation that the difference in 'inner' and 'outer' C-C distances becomes much smaller on coordination.

Answers

5.1 **5.18** shows molecular orbital diagrams for the two geometries. Notice that for both the octahedral and square planar geometry all of the ligand σ orbital combinations find a symmetry match with a central atom orbital (contrast this with the case in Question 5.4) and that both arrangements lead to a good *HOMO-LUMO* gap, for low spin d^6 and d^8 respectively. If the ligand properties are such that the *HOMO-LUMO* gap is large, then the orbital structure leads to predictions of stable 18 and 16 electron compounds respectively. Notice though, that there is one orbital, metal p_z in the square

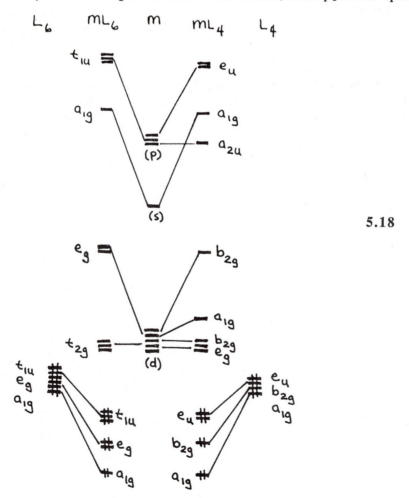

5.18

planar case which, *by symmetry*, cannot interact with any ligand combination. This distinguishes the two geometries electronically. The p_z orbital lies too high in energy to be filled with electrons (unlike the xz, yz and xy orbitals which also cannot interact in a σ fashion with the ligands) and thus the maximum stable electron count for the square planar geometry is 16. $V(CO)_6$, $Cr(CO)_6$ and $Fe(CO)_6$ are 17, 18 and 19 electron species respectively. $Fe(CO)_6$ violates the 18 electron rule, and should not be stable. It's instability arises *via* population of one of the e_g pair of orbitals, antibonding between metal and ligand. A large *HOMO-LUMO* gap here signals a strongly antibonding e_g pair and a ligand is lost (in this case to give the stable $Fe(CO)_5$ molecule). $V(CO)_6$ has only 17 electrons and readily captures one more to give the stable $V(CO)_6{}^-$ species. It is stable in the sense that a ligand is not spontaneously lost, a result easily understandable from the molecular orbital diagram since no antibonding orbital is occupied.

5.2   These are shown in **5.19**, and each, of course, represent only one of a myriad of selections.

5.3   Given the caveats that the number of *CO* groups are different in the molecules to be compared, that the geometries of the three molecules are different, and that comparisons of vibrational data should be made with force constants rather than frequencies, the result that *CO* stretching frequencies increase on moving from left to right across the transition metal series is a general one. **5.23** shows the interaction diagram appropriate for the interaction of metal and *CO*. Of crucial importance is the energy separation

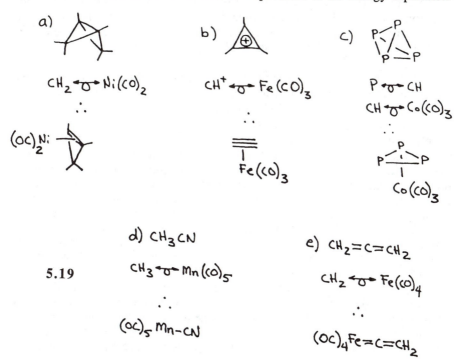

between the metal d levels and the π^* orbitals of CO. Interaction of these two orbitals leads to backdonation of electron density from metal to CO. In orbital terms it leads to population of the CO π^* orbital and hence a reduction in vibrational frequency. The magnitude of the interaction decreases as the metal d/CO π^* energy separation increases. As the transition metal series is traversed from left to right the metal d levels drop in energy (by the time the main group elements are reached they have become core orbitals). Thus this crucial energy gap increases, backdonation decreases and the vibrational frequency increases on moving across the series. Another way to change the vibrational frequency is described in Question 3.8.

5.4   Within the C_{3v} point group the three 'radial' π orbitals (**5.20**) transform as $a_1 + e$ and the three 'tangential' orbitals as $a_2 + e$. In this point

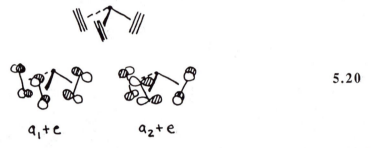

5.20

a_1+e a_2+e

group the metal d orbitals transform as $a_1 + 2e$, the s orbital as a_1 and the three p orbitals as $a_1 + e$. Thus there is no symmetry match between one ligand combination (a_2) and a metal orbital. This ligand combination remains non-bonding, and the two electrons in it ignored in electron counting considerations. Thus the 'real' electron count in this complex is $6(W) + 2(CO)$ $+ 12$ ($3 \times C_2H_2$) - 2(nonbonding a_2) = 18.

5.5.   **5.21** shows the construction of a molecular orbital diagram for this molecule. There is a stabilization of the H_2 σ orbital with a concurrent destabilization of the a_1 $s/z^2/z$ hybrid orbital which points out of the bottom of the square pyramid. A good *HOMO-LUMO* gap is opened up on coordination, appropriately for what is now an 18 electron molecule. The H_2 σ^* orbital interacts with one of the occupied π-type d orbitals, leading to a stabilization. The interaction is therefore very similar to the interaction of olefins with transition metals, but with one striking difference. In the olefin case there is an underlying σ framework to hold the two carbon atoms together, but here there is solely the H-$H\sigma$ bond. If backdonation into the H-H σ^* orbital is extensive then the H-H bond might be expected to break resulting in dihydride formation. Thus control of the extent of this interaction enables selection of the classical or nonclassical structure. Since the interaction energy is proportional to $S^2/\Delta E$, where S is the overlap integral and ΔE their energy separation, we can vary either parameter. One way to reduce the extent of such backdonation is to lower the energy of the metal d orbitals and thus increase ΔE. This may be achieved by ensuring that a positive charge resides on the complex. Recall a similar effect in metal carbonyl chemistry (Question 3.8) where the CO stretching frequencies of one isoelectronic series decrease in the order

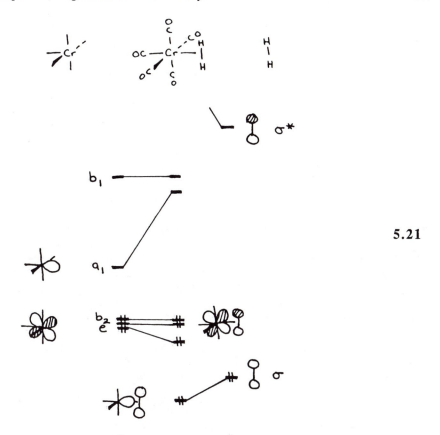

5.21

$Mn(CO)_6^+ > Cr(CO)_6 > V(CO)_6^-$. The complex with the positive charge has the deepest-lying metal d levels, therefore the smallest backdonation interaction and thus the highest CO stretching frequency. A second route involves a reduction in S. By attaching strong π-acceptor ligands to the complex, the d orbital character in the metal b_2 orbital is reduced such that overlap with the H-H σ^* orbital is lowered.

5.6. (a) $V(CO)_6$ with five electrons is Jahn-Teller unstable in an octahedral geometry (a $^2T_{2g}$ electronic state) using the molecular orbital diagram of **5.18**. The structure of the $V(CO)_6$ molecule, as determined by X-ray crystallography reflects this. The molecule has a small distortion away from octahedral. Since the Jahn-Teller instability is associated with occupation of π-type orbitals, generally regarded as less stereochemically potent than σ-type orbitals, the driving force away from the symmetrical structure is smaller than in d^9 chemistry for example. The result is a dynamic rather than static distortion. The origin of the broader absorbion band is shown in **5.22**, where the vibrational motion has a considerably larger amplitude than for the Jahn-Teller stable $Cr(CO)_6$ molecule. (b) A low-spin d^8 molecule in an octahedral geometry gives rise to $^1A_{1g}$ and ^1E_g electronic states. The ^1E_g state lies lower in energy and is Jahn-Teller unstable. The Jahn-Teller active coordinate for

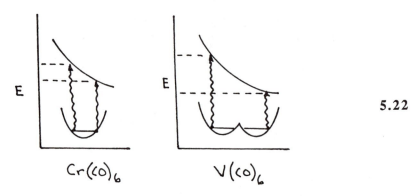

$Cr(CO)_6$ $V(CO)_6$

5.22

this state is given by the symmetric direct product of e_g, namely $a_{1g} + e_g$. A motion of e_g symmetry leads to loss of two axial ligands and generation of the square plane. For further discussion see Question 3.11). (c) The metal based e_g orbitals of the octahedral complex (5.18) are strongly metal-ligand antibonding for the case of CO. Their population on photolysis leads to ligand loss. In solution with ligands L this process gives rise to a useful synthetic route to $Cr(CO)_5L$ molecules, but when storing the material, exclusion of UV light improves its shelf life.

5.7   The stability of $X_2 = CO$ and N_2 complexes of transition metals arises *via* the two interactions shown in 5.23. Interaction of the filled X_2 σ orbital with an empty metal orbital leads to donation of electron density to the metal, and interaction of the empty X_2 π^* level with a filled metal d orbital, the reverse. The strength of the interaction depends on the ratio of $S^2/\Delta E$, where S is the overlap integral of the interacting orbitals and ΔE their energy separation. The relative stability of the M-CO, M-NN and M-OC linkages, decreasing in this order, are readily understood from this model by considering the nature of the σ and π^* levels of the two ligands (5.24). For the CO ligand notice that the π^* level is largely carbon located, a direct result of the carbon/oxygen electronegativity difference. In N_2 the π^* orbital is localized equally on the two nitrogen atoms by symmetry. Thus the strength of the metal-π^* interaction should decrease in the order $\underline{C}O > N_2 > \underline{O}C$. The nature of the σ orbital in CO is complex to derive. It is a very weakly antibonding orbital but is largely lone-pair in character. Importantly it is localized on carbon. In N_2 this lone pair character is preserved but is equally distributed on the two nitrogen atoms. Thus the strength of the σ -donor interaction is expected to decrease in the order $\underline{C}O > N_2 > \underline{O}C$. Thus both interactions decrease in the same order. This shows up in the larger number of stable CO complexes compared to N_2 ones, the larger frequency shifts compared to the free ligand for carbonyls compared to N_2 complexes and the non-existence of the simple M-OC linkage. (M-OC-M' linkages are known.) One place to look for an isocarbonyl might be with electropositive atoms (*e.g.*, Sc and the lanthanides) but these metals are so oxophilic that a CO molecule stands a good chance of being destroyed in the process.

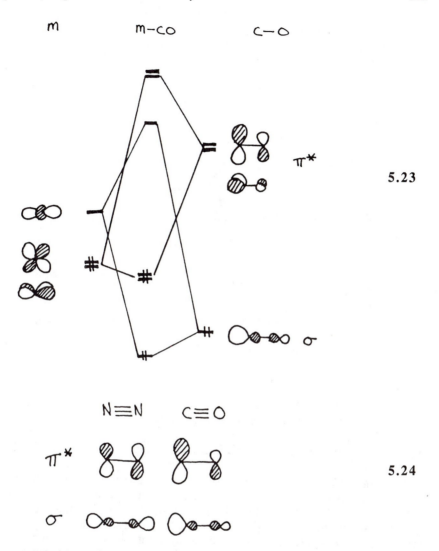

5.23

5.24

5.8 **5.25** shows the assembly of the molecular orbital diagram. The strongest interaction occurs for the *e'* orbitals. The *As₂ π* orbitals point perpendicular to the *As-As* axis and overlap here with the metal orbitals is quite good. Although by symmetry the *[Cr(CO)₅]₃ a₁'* combination may interact with the *As₂* orbitals of the same symmetry, the energetically closest one is a lone pair orbital which points away from the the *As-As* bond and so this interaction may be small. We have left out the antisymmetric combination of lone pair orbitals on *As₂* in **5.25**. If we regard the two pairs of electrons in the *1e'* orbitals as being associated largely with *As-Cr* bonding rather than *As-As* bonding then the *As-As* bond order has dropped to unity. Of course such divisions are always somewhat arbitrary. In fact the *As₂ σ* bond can and will overlap to some extent with the *a₁* combination which will reduce the As-As bond order. Furthermore, there is a combination of occupied *Cr d* centered

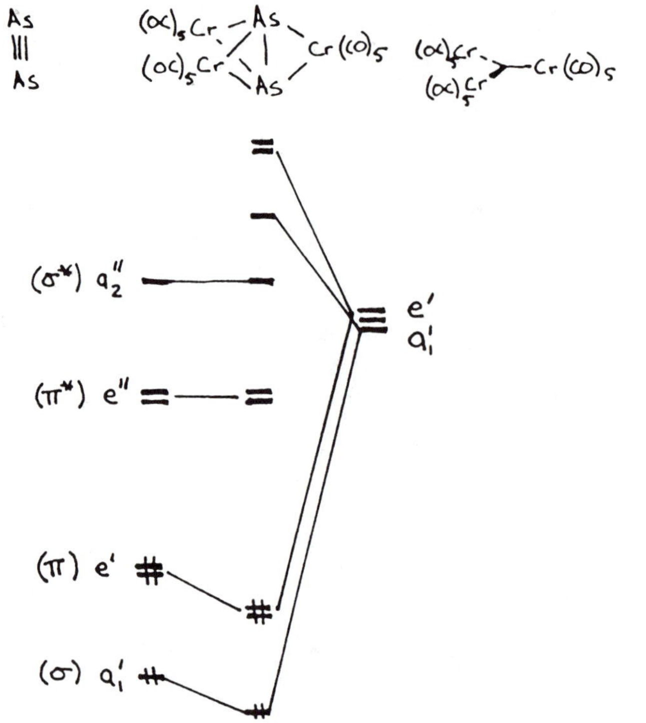

5.25

orbitals of e'' symmetry which will overlap with and partially occupy the π^* set in As_2. This will reduce the bond order further.

5.9   The σ frameworks of **a** and **b** are those appropriate for a d^{10} trigonal planar ML_3 and low-spin d^8 square planar species respectively, and are shown in **5.26**. In each case we are interested in the interaction of the ethylene π^* orbital with this σ set in the two conformations. For the in-plane conformation the ethylene π^* orbital can interact with the xy and $x^2 - y^2$ orbitals and for the out-of-plane conformation with the xz and yz orbitals. On simple energy gap arguments we can see that for **a** the stronger interaction is for the in-plane conformation. The in-plane e' pair of orbitals in D_{3h} are σ antibonding and are pushed to higher energy than the out-of-plane nonbonding e'' pair, *i.e.*, the observed geometry is predicted to be the more stable one. In **b** the highest-lying orbital, $x^2 - y^2$, is unoccupied and the other three d orbitals lie close together. Including π donor interactions with the other three atoms, xy , with three π antibonding interactions, lies above the xz, yz pair with only one or two. As a result we expect the electronic contribution to the conformational energy difference to be smaller than for **a**. The in-plane conformation however, experiences quite a serious steric repulsion, and this is certainly the reason for the adoption of this structure.

5.10. **5.27** shows the assembly of a molecular orbital diagram for this molecule. There is a stabilization of the lowest energy (a_1) H_3 orbital with a

a   b

z

x ⊥ y

— b_{1g} (x^2-y^2)

(x^2-y^2, xy) e' ⧣

5.26

(z^2) a_1' ⧺   ⧺ a_{1g} (z^2)

(xz, yz) e'' ⧣   ⧣ b_{2g} (xy)
                                e_g (xz, yz)

concurrent destabilization of the a_1 $s/z^2/z$ hybrid orbital which points out of the bottom of the square pyramid. The antibonding e' orbitals of H_3 find a match with the e pair of orbitals of the transition metal fragment, although the degeneracy is lost through the reduction in symmetry. Thus there is both donation to the metal (a_1 orbitals) and backdonation too (e orbitals). A good *HOMO-LUMO* gap appears at the d^6 electron count and an 18 electron molecule is generated. This would correspond to the neutral molecule $V(CO)_5(H_3)$. The orbital interaction diagram is of course very similar to the one which would be generated for the molecule $V(CO)_5(C_3H_3)$ containing a cyclopropenium unit. Just as Hückel's rule tells us that such a moiety should be stable formally as $C_3H_3^+$, similar arguments should apply to the stability of H_3^+. What is different, and of crucial importance, is that the $C_3H_3^+$ system will hold together *via* the underlying σ framework when the π* orbitals are partially occupied by interaction with the two filled members of the 't_{2g}' set in $V(CO)_5^-$. This is not the case for H_3^+; significant occupation of the e' set will significantly weaken and destroy H-H bonding. Thus the successful isolation of an intact H_3^+ ligand will occur only if π-backdonation from t_{2g} to e' is minimized and will thus be senstive to the details of the d^6 $M(CO)_5$ fragment. A $Mn(CO)_5^+$ or $Cr(CO)_5$ species would be better than $V(CO)_5^-$ in this regard since the t_{2g} set in the former will lie deeper in energy (from electronegativity and charge reasons) and thus not interact as strongly with the e' set as in $V(CO)_5^-$. See question 5.5 for a related problem.

5.11. This result is not really too surprising in fact. Each $Pt(PR_3)_2$ unit is isolobal with CH_2, and thus the dimer $(PR_3)_2Pt-Pt(PR_3)_2$ corresponds to the ethylene molecule with a double bond between the heavy atoms. In orbital terms it is a little more complex. **5.28a** shows one possible way of generating the molecular orbital diagram for the dimer. Only degenerate orbital interactions are included, and the result is the one expected for repulsion between two closed shells. **5.28b** shows a second possibility which highlights

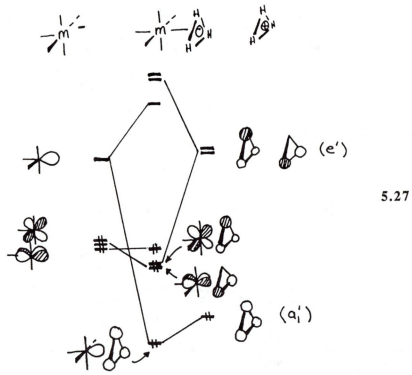

5.27

the isolobal relationship. There is now one σ and one π bond between the metal atoms. In fact from calculation (*J. Amer. Chem. Soc.*, **100**, 2074, (1978)) the situation is neither of these two. The σ orbitals of the two types strongly mix together (**5.29**) to further strengthen the bonding of the deepest-lying σ

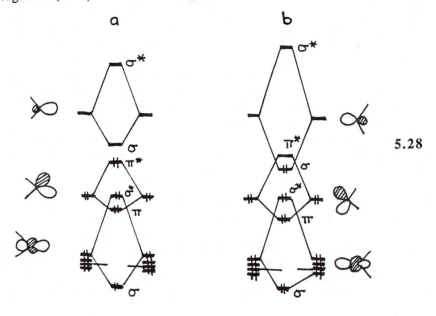

5.28

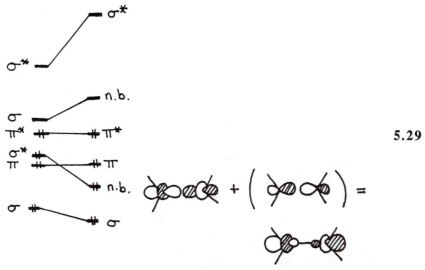

5.29

orbital, and the σ^* orbitals mix together such that the lower of the two is now best described as a nonbonding orbital.

5.12. **5.30** shows the assembly of the molecular orbital diagram for octahedral H_6 and the effect of adding an interstitial carbon atom. The octahedral H_6 orbitals are easy to derive using group theory. Their energetic ordering, $a_{1g} < t_{1u} < e_g$, is set simply by the number of the bonding and antibonding interactions along the octahedral edges. Interaction with the $2s$ and $2p$ orbitals of carbon leads to the generation of four deep-lying orbitals, bonding between carbon and hydrogen and either bonding or non-bonding between hydrogen atoms. A total of eight electrons is needed for stability, a number exactly furnished by the molecule in question.

5.13. (a) First we need to form the relevant symmetry adapted linear combinations of the frontier orbitals of the NH unit shown in **5.31**. The assembly of the molecular orbital diagram of the molecule is shown in **5.32**. The form of the orbitals is particularly simple. One potentially complicating feature is removed if you notice that the overlap of $1e'$ on $(NH)_3^{6-}$ with the x,y set on Os^{6+} is large while that between $1e'$ and $x^2 - y^2$, xy is small. Likewise, the overlap of $2e'$ on $(NH)_3^{6-}$ with $x^2 - y^2$, xy is large but that between $2e'$ and x,y is small. The form of the d-centered orbitals is shown in **5.33**. (b) Recall

$$N{-}H^{2-}$$

5.31

H_6^{2+} CH_6^{2+} C

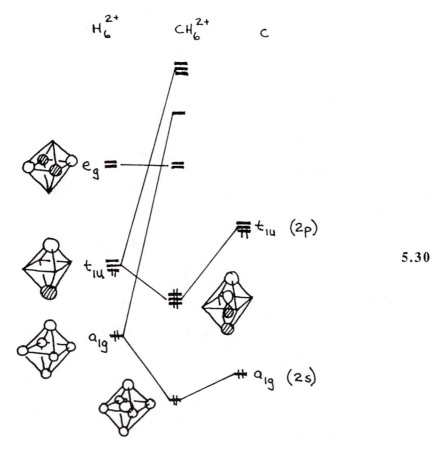

5.30

that only the σ electrons are included for counting purposes. Consequently each NR^{2-} unit donates two electrons and since this is a d^2 complex, this gives a total of eight electrons. An eight electron ML_3 complex ought to be pyramidal but this one is planar. The *HOMO* we might expect to be stabilized on bending is shown in **5.34**, and in fact is from a calculation. However there is strong π bonding between the *NH* units and the *Os* atom. The molecular orbitals *2e'*, *1a2"* and *1e"* orbitals all lose π overlap and are therefore destabilized on pyramidalization. A related problem turns up in Question 3.27. Counting the molecule as a 20 electron system means that each NR^{2-} unit donates six electrons as in **5.35**. In fact this is not correct. As the interaction diagram shows the a_2' orbital cannot interact by symmetry with any atomic orbital on the metal. (A similar situation arises in Question 5.4.) Thus the electron count is only 18. But this is still two too many; planar ML_3 complexes should have 16 electrons. The 16 electron rule comes about by leaving the metal z orbital uninvolved in bonding, but here we can see that there is strong interaction with the $a_2"$ combination on the ligands. This is the identity of the extra two electrons.

5.14. The answers are shown in **5.36**.

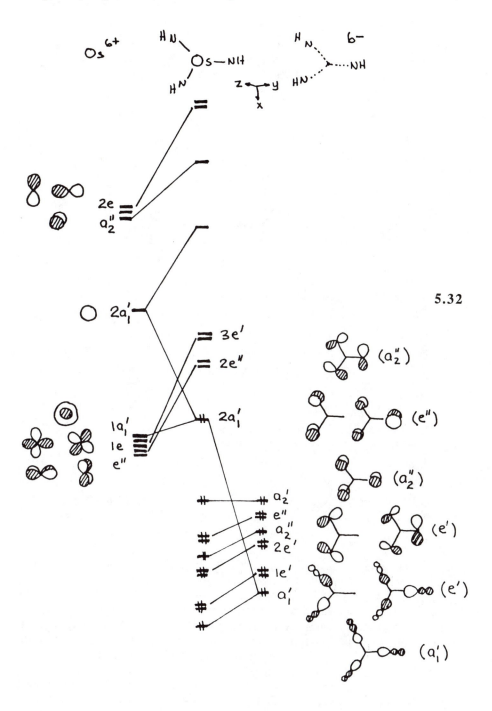

5.32

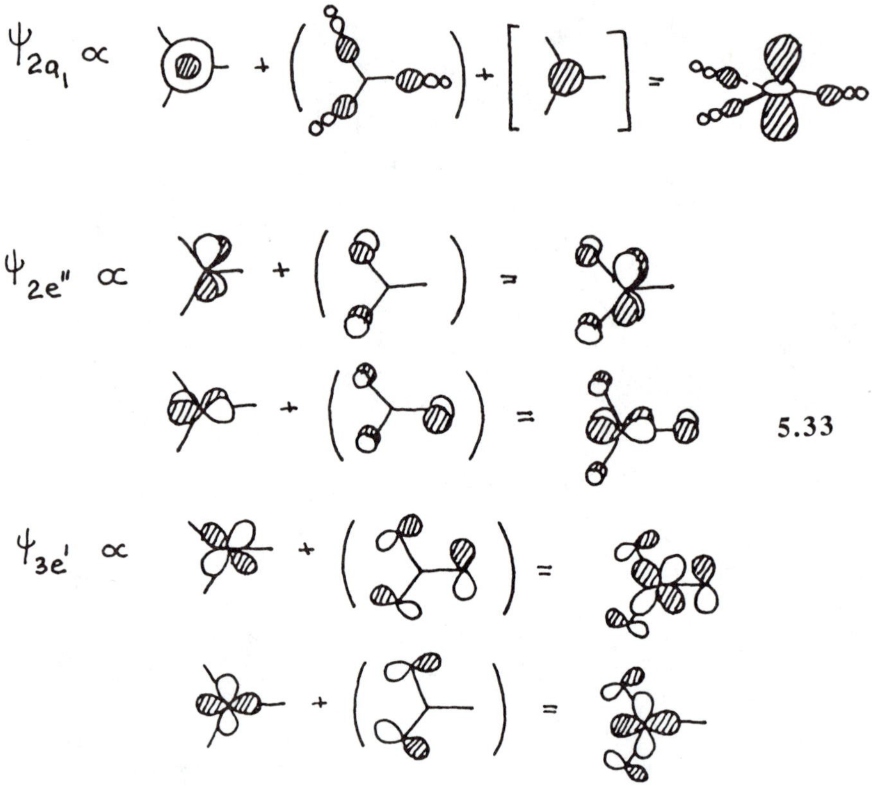

$\psi_{2a_1} \propto$ ... $=$ ...

$\psi_{2e''} \propto$ ... $=$ ...

... $=$ **5.33**

$\psi_{3e'} \propto$ ... $=$ ...

... $=$ ...

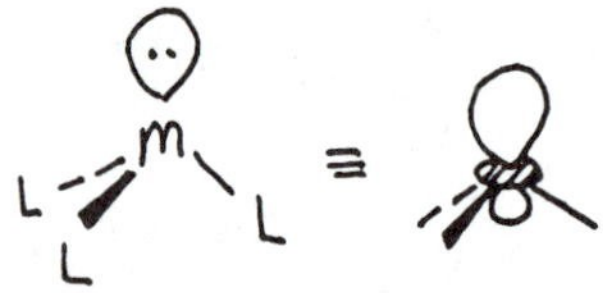

$=$ **5.34**

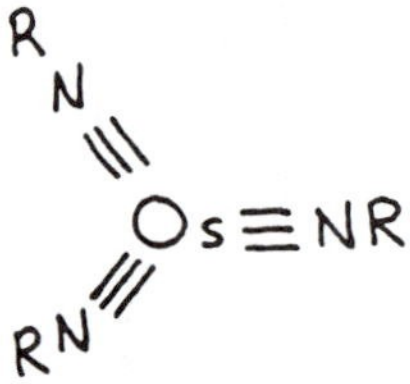

5.35

5.15. A molecular orbital diagram is shown in **5.37** where the T_d point group is used for simplicity. We have used an sp^3 hybrid orbital on each methyl group, and since carbon is considerably more electronegative than lithium, have only used the valence $2s$ orbitals on Li. The σ^* levels of CH_3 are probably too high in energy to be important and are left off the diagram. The $a_1 + t_2$ methyl orbital combinations find a perfect symmetry match with the corresponding orbitals of the Li_4 unit. Four bonding orbitals result, which, when filled with electrons match the actual electron count in the molecule. These orbitals are bonding between the lithium and carbon atoms, but overall are nonbonding between the lithium atoms. This may be seen by inspection of their nodal characteristics at the left hand side of the diagram. While the a_1 orbital is Li-Li bonding, the t_2 orbital is Li-Li antibonding. Thus the molecule is held together by Li-Me bonding. Compare this result with that of Question

a) $(CO)_4 Re^{2-}$ ⟷σ⟶ CH_2^-

     $(CO)_4 Re$ ⟷σ⟶ CH_2^+

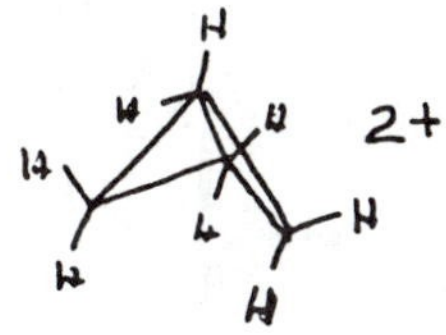

b) Ph_3P Pt ⟷σ⟶ CH_2 (with OC)

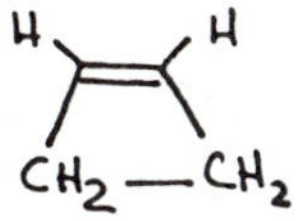

c) $CpNi$ ⟷σ⟶ CH

     $(CO)_3Co$ ⟷σ⟶ CH

     $(CO)_3Fe$ ⟷σ⟶ CH^+

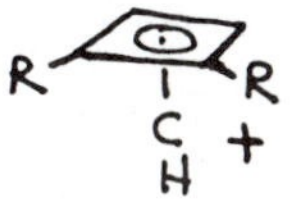

5.36

d) $Re(CO)_5$ ⟷σ⟶ CH_3

     $Pt(CO)_2$ ⟷σ⟶ CH_2

$CH_3-CH_2-CH_3$

e) $CpRh(CO)$ ⟷σ⟶ CH_2

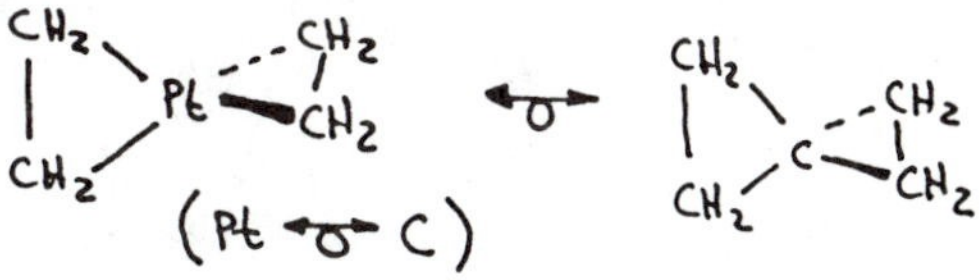

$(Pt$ ⟷σ⟶ $C)$

f) $CpW(CO)_2$ ⟷σ⟶ CH

     $(Me_3P)_2Pt$ ⟷σ⟶ CH_2

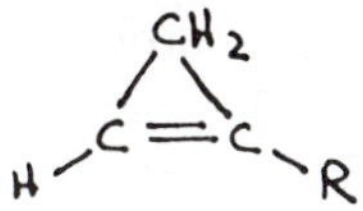

g) $(CO)_3Co$ ⟷σ⟶ CH

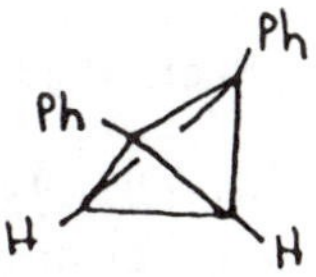

h) $CpRh(CO)$ ⟷σ⟶ CH_2

     $Mo(CO)_5$ ⟷σ⟶ CH_3^+

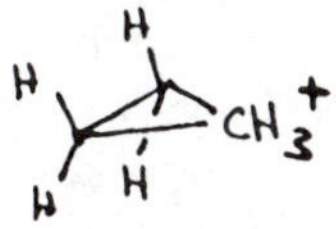

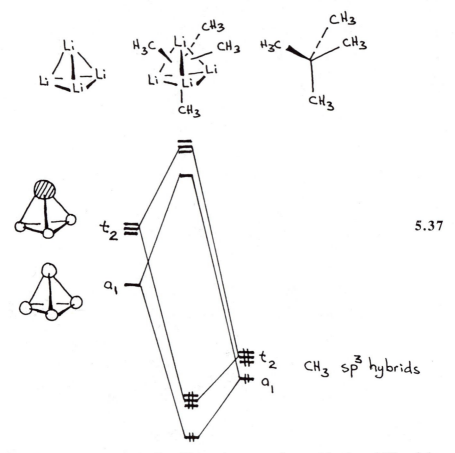

5.37

5.12. The two are very similar. The carbon atom located in the middle of the $(AuPR_3)_6$ cluster plays the same orbital rôle as the bridging methyl groups do here.

5.16. (a) $Ni_5(CO)_{12}{}^{2+}$ = $[Ni(CO)_2]_3[Ni(CO)_3]_2{}^{2+}$. Skeletal electron count = 6 $(3xNi(CO)_2)$ + 8$(2xNi(CO)_3)$ - 2(2^+) = 12. *closo* trigonal bipyramid.

(b) B_5H_9 = $(BH)(BH_2)_4$. Skeletal electron count = 2(BH) + 12$(4xBH_2)$ = 14. *nido* octahedron.

(c) $H_2Ru_6(CO)_{18}$ = $[Ru(CO)_3]_6{}^{2-}$. Skeletal electron count = 12 $(6xRu(CO)_3)$ + 2$(2-)$ = 14. *closo* octahedron.

(d) $Fe_5(CO)_{15}C$ = $[Fe(CO)_3]_5{}^{4-}$. Skeletal electron count = 10 $(5xFe(CO)_3)$ + 4$(4-)$ = 14. *closo* octahedron with 'interstitial' carbon. Alternatively: $Fe_5(CO)_{15}C$ = $[Fe(CO)_3]_5C$. Skeletal electron count = 10 $(5xFe(CO)_3)$ + 2(C) = 12. With n skeletal atoms and n skeletal electron pairs a capped structure is often found. In this case the structure will be a capped trigonal bipyramid.

(e) $(CpCo)_2C_2B_6H_8$ = $(CpCo)_2[CH]_2[BH]_6$. Skeletal electron count = 4 $(2xCpCo)$ + 6$(2xCH)$ + 12$(6xBH)$ = 22. *closo* bicapped square antiprism.

f) B_9H_{15} = $(BH)_3(BH_2)_6$ Skeletal electron count = 6$(3xBH)$ +

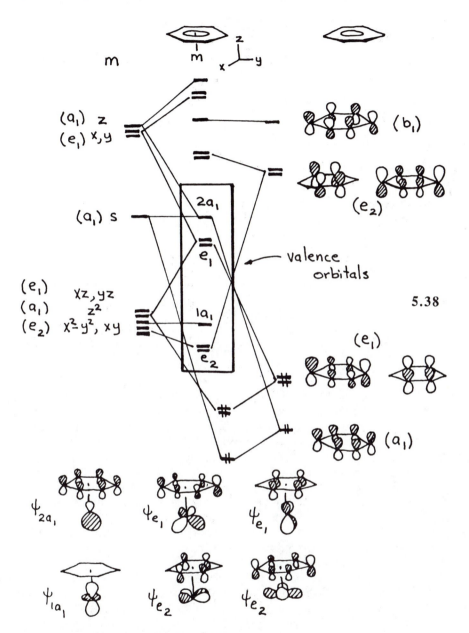

$18(6xBH_2) = 24$. *arachno* undecahedron.

g) $Co_8C(CO)_{18}^2 = C[Co(CO)_2]_6[Co(CO)_3]_2^{2-} = [Co(CO)_2]_6$
$[Co(CO)_3]_2^{6-}$. Skeletal electron count $= 6(6xCo(CO)_2) + 6(2xCo(CO)_3) + 6$
$(6^-) = 18$. *closo* dodecahedron with an interstitial carbon atom. A similar
question concerning the carbon atom noted in (d) arises here too.

h) $(PhC\equiv CPh)Ru_4(CO)_{12} = [CPh]_2[Ru(CO)_3]_4$. Skeletal electron count
$= 6(2xCPh) + 8(4xRu(CO)_3) = 14$. *closo* octahedron.

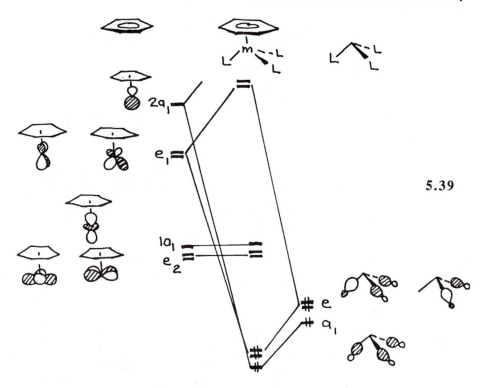

5.39

5.17. (a) The assembly of the (η^6-benzene)M orbital diagram is shown in
5.38. Notice that the nodal plane of z^2 is nearly coincident with the p atomic
orbitals of carbon, and thus that it is left nonbonding. The electronic structure
of (η^6-benzene)ML_3 is shown in **5.39**. The ultimate electronic structure for
η^6-benzene-ML_3, as well as, ferrocene (see question 5.23) and octahedral ML_6
(question 3.1) has six ligand-based orbitals that are at low energy and are
always filled. There are always three non-bonding, metal based orbitals, also
filled, and at higher energy two empty metal-ligand antibonding orbitals.

5.18. **5.40** shows the frontier orbitals of $NiCp^+$ (it is isolobal with $Fe(CO)_3$
and (η^6-benzene)Fe in the answer to question 5.17a) and their interaction with
the orbitals of two sulfur atoms of the ligand. The metal-ligand σ interaction
leads to a destabilization of the xz orbital leading to a significant gap between
yz and xz and thus a diamagnetic complex. When the π interaction is switched
on then yz is pushed up in energy and can end up close to xz. With a small gap
between these orbitals then high-spin (triplet) molecules are possible. The
energetic balance between paramagnetic and diamagnetic molecules is
determined in part by the balance betweeen these σ and π interactions.

5.19. The Dewar-Chatt-Duncanson model for olefin coordination to transition
metals involves σ donation from the filled olefin π orbital to the metal and
back-donation from a filled metal level to π^*. The orbital details are
considered in Question 5.21. Such synergistic bonding is characteristic of the
coordination of many ligands (e.g., $\eta^2 H_2$, Cp, benzene, olefins, CO, PR_3 etc.)

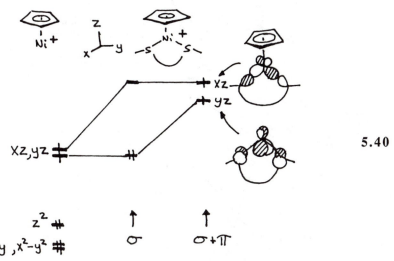

5.40

$$z^2 \uparrow\downarrow \qquad\qquad \uparrow \qquad\qquad \uparrow$$
$$xy, x^2-y^2 \uparrow\downarrow\uparrow \qquad\qquad \sigma \qquad\qquad \sigma+\pi$$

to a metal. But what is the relative importance of the two parts of the process? Since **5.2** shows a strong correlation with the energy of the *LUMO* (π^* level) of the olefin, this interaction is probably dominant here.

5.20. The orbitals χ_1 to χ_4 of **5.8** transform as shown in Table 5.1 and normalized symmetry adapted linear combinations are shown in **5.41**.

Table 5.1

D_{2d}	E	$2S_4$	C_2	$2C_2'$	$2\sigma_d$	
Γ	4	0	0	0	2	$a_1 + b_2 + e$

$$\psi_{a_1} = \frac{1}{\sqrt{4+4S_1+8S_2}}(\chi_1+\chi_2+\chi_3+\chi_4) \equiv$$

$$\psi_{b_2} = \frac{1}{\sqrt{4+4S_1-8S_2}}(\chi_1+\chi_2-\chi_3-\chi_4) \equiv$$

5.41

$$\psi_e \quad = \frac{1}{\sqrt{2-2S_1}}(\chi_1-\chi_2) \equiv$$

$$\quad = \frac{1}{\sqrt{2-2S_1}}(\chi_3-\chi_4) \equiv$$

where $S_1 = \langle \chi_1 | \chi_2 \rangle$, etc.
$S_2 = \langle \chi_1 | \chi_3 \rangle$, etc.

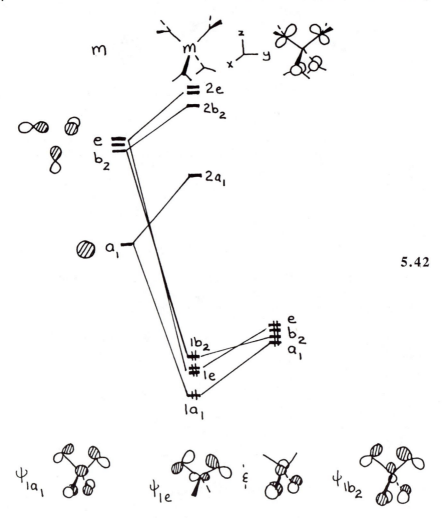

The molecular orbital diagram is assembled in **5.42**.

5.21. A generic metal-olefin interaction energy diagram is shown in **5.43**. It shows both a ligand-to-metal donation interaction (controlled by the size of ΔE_1) and a backdonation interaction (controlled by the size of ΔE_2). Attaching electron-withdrawing substituents to the olefin (resulting in an increase in the sum of the Hammett constants for the ligands) leads to an increase in the positive charge at carbon and a drop in the energy of the π and π^* levels of the olefin (*i.e.*, the π ionization potential increases). As a result the interaction between the filled metal and empty π^* levels increases, but that between the filled π and empty metal levels decreases. The plots of **5.9** show two different trends. Since for the nickel complexes the stability constant increases (*i.e.*, the olefin complex becomes more stable) with increasing $\Sigma\sigma$ we conclude that the backdonation interaction is the more important of the two, the same result obtained by consideration of the variation in the olefin *LUMO* energies in

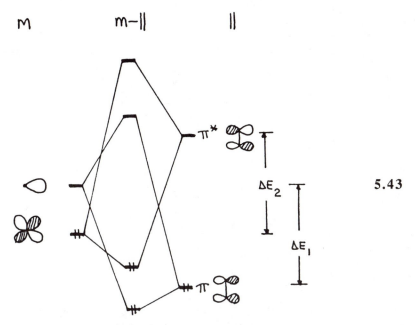

Question 5.21. The behavior of the stability constants of the silver complexes is opposite to that for the nickel ones. Electronically their stability is dominated by variation ΔE_1. This is relatively easy to understand. On electronegativivity grounds the filled (d^{10}) set of orbitals for Ag^+ lie deep in energy and are quite contracted. The stronger interaction is thus between the empty p orbitals on silver with the olefin π orbital. For the nickel case (Question 5.19.) it is difficult to predict *a priori* how the plot would have turned out. Both interactions are probably important but that involving π^* dominant. (*J. Amer. Chem. Soc.*, **96**, 2780 (1974))

5.22. **5.44** shows molecular orbital diagrams for the two conformations which show important differences in the interaction between the ethylene π^* orbital and the metal d orbitals. In the middle of the diagram is the splitting pattern for a trigonal bipyramid. *i.e.*, we have considered the ethylene π orbital to be approximately as strong a σ donor as the carbonyl σ orbital. Next we have switched on π interactions between the olefin (π^*) and the metal orbitals. In the perpendicular conformation this interaction involves the metal e'' orbitals (using D_{3h} labels) and in the in-plane conformation the e' orbitals. Since the orbital interaction energy varies inversely with the energy separation of the two orbitals concerned, the in-plane conformation is predicted to be the more stable of the two (i.e., $|\Delta E_1| > |\Delta E_2|$). Furthermore, there is an important overlap difference which reinforces this conclusion. The e' set contains some metal p character causing the xy component to be hybridized towards the π^* orbital wheareas this does not occur in the e'' set.

5.23. (a) This is a much simpler proposition than appears at first sight. At the right hand side of **5.45** are the Cp π orbitals drawn in a way which highlights their nodal properties. Ascending the stack of π levels (see Chapter 1.11) is

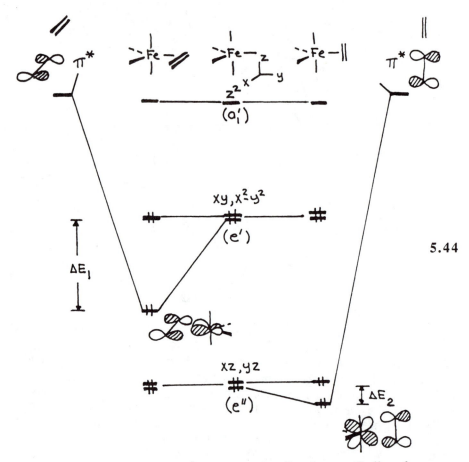

5.44

associated with an increasing number of nodes. For the two *Cp* ligands we may simply write in- and out-of-phase combinations of the individual π orbitals. Now all that remains to be done is to find the relevant symmetry matches with the metal orbitals and drawing out the orbital diagram. This can be done in a purely qualitative way. Thus the a_{2u} orbital clearly has the same nodal properties as z, the pair of e_{1u} orbitals the same as x,y, and the e_{1g} combination the same as those of xz and yx. The pair of orbitals of e_{2g} symmetry have the same characteristics as the xy and $x^2 - y^2$ pair, but e_{2u} finds no symmetry match. The sizes of the interactions we draw reflect overlap considerations. Thus the xy and $x^2 - y^2$ pair are stabilized by only a small amount due to the δ–type overlap with the ligand orbitals. There is one three-orbital interaction. Clearly s, z^2 and a_{1g} of the Cp_2^{2-} set have the same symmetry. For the same reason given in the asnwer to question 5.17, the a_{1g} combination has a small overlap with z^2. Consequently it is left nonbonding and the primary interaction occurs with metal s. Notice that there are (as in question 5.17 for benzene$Cr(CO)_3$) three metal d orbitals lying low in energy and two d orbitals strongly antibonding to the *Cp* ligands at higher energy. The octahedral splitting pattern is once again restored! *A priori* it is difficult to predict the

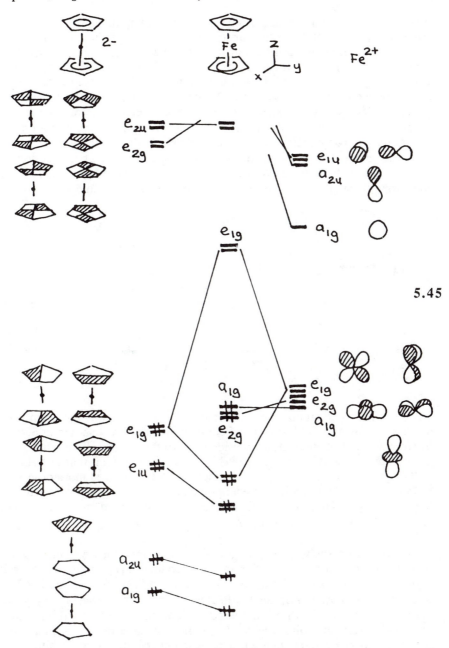

5.45

a_{1g}/e_g level ordering. (b) An eighteen electron count arises for the molecule when all the levels are filled through 'z^2', of which twelve occupy strongly bonding orbitals (thus each ring is held by three pairs of electrons) and six occupy nonbonding or weakly bonding orbitals. (c) With two extra electrons the xz,yz pair are occupied and a triplet results. (d) The existence of the nickelocene molecule indicates that the xz,yz pair of orbitals lie low in energy.

It is then perhaps not too surprising that their population can be controlled by the nature of the metal and the substituents on the Cp ligand. This is the case with manganocene. Here high and low spin variants are possible. Since the xz,yz pair of orbitals are metal-Cp antibonding their population is expected to lead to longer bond lengths. This is indeed the case. The hextuplet form (**5.46**) has a longer bond than the doublet. But why is Cp_2Mn high spin while $(Me_5C_5)_2Mn$ is low spin? A rationale can be offered in terms of the donor properties of the methyl substituent. Such a methyl group is weakly π donating so that the e_{1g} combination of Cp_2^{2-} orbitals lies higher in energy and closer to the metal d set than in the unsubstituted parent. A stronger interaction with metal xz and yz then will lead to a destabilization of the molecular e_{1g} orbital, increasing the gap to the molecular a_{1g}/e_{2g} set as indicated in **5.46**.

5.46

5.24. (a) The σ diagram is constructed as in **5.47** and (b) the form of the orbitals is shown in **5.48**. (c) The effect of π interactions is shown in **5.49**. (d) In structure **b** all σ interactions are identical to those in **a**. What changes are the π interactions as shown in **5.50**. Now it is the e'' orbital that is stabilized. So the question to be answered is whether e' in **a** is stabilized more or less than e'' in **b**. Since e' lies at a higher energy than e'', the energy gap between e' and the p orbitals on SnH_2 is smaller. As a result e' is stabilized more than e'' and **a** becomes the favored conformation.

5.25. For both cases we can start off with the σ-only square planar splitting pattern and then add in the effect of the π-bonded CH_2 groups. The orbital diagrams for the two conformations are shown in **5.51**. Which one lies lower in energy, i.e., how is $2\Delta_1$ related to Δ_2? We need to answer a very general chemical question. If there are n equivalent central atom orbitals and n equivalent ligand σ orbitals, which is the more stable arrangement, that where the n ligand σ orbitals interact with just one central atom orbital (as in conformation **a**) or the n ligand orbitals interact with each of the central atom orbitals. As shown in Question 3.18 the second is the more stable for the case where there are suficient electrons to just fill all of the bonding orbitals which result. A similar result (Question 4.14) is behind the stabilization of the tetrahedral geometry (all three orbitals used equally) over the square planar geometry (one orbital not used at all). In other words we need to minimize the use of 'busy' and 'idle' orbitals, and make sure they are used equally.

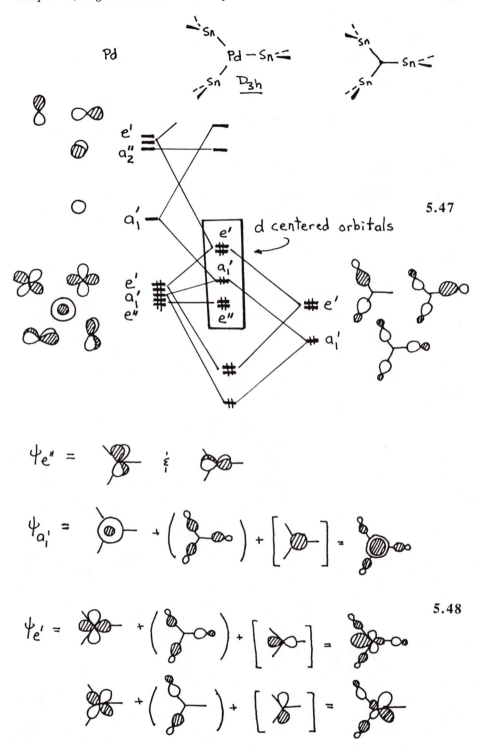

5.47

d centered orbitals

5.48

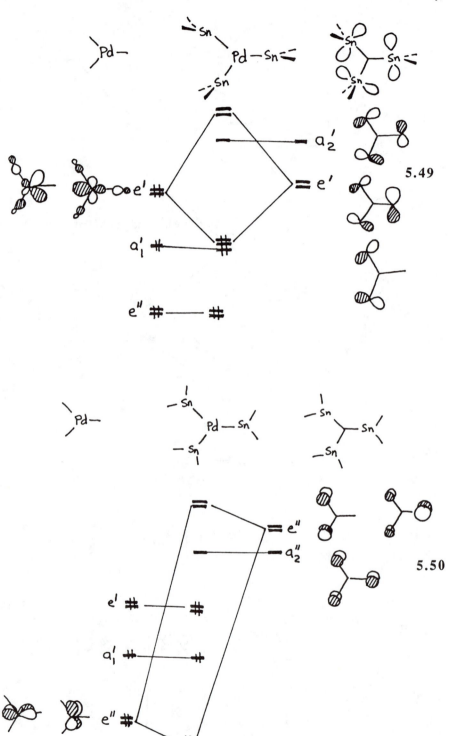

5.49

5.50

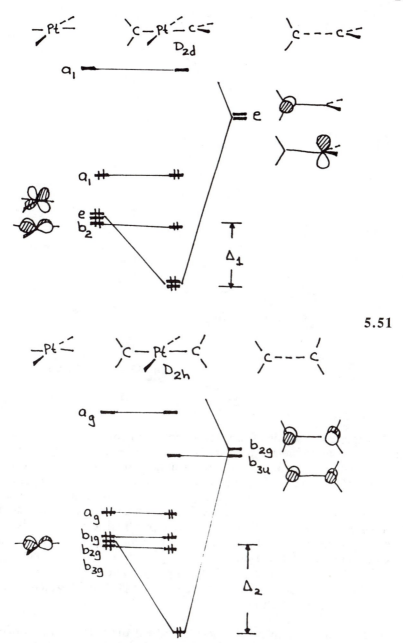

5.51

5.26. **5.52** shows the assembly of a molecular orbital diagram for the linear molecule. There are three orbitals which transform as σ_g^+, which leads to one bonding orbital (ligand centered), one antibonding orbital (largely *s*) and one nonbonding orbital (largely metal *d*). So in this geometry there are are two bonding orbitals followed by five nonbonding orbitals giving a total of seven orbitals which can accommodate a total of fourteen electrons as shown in

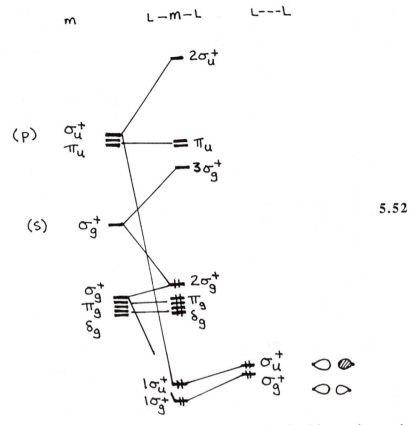

5.52. The metal x,y pair of orbitals do not get involved in any interactions at this geometry and are too high in energy to be filled. **5.53** shows the Walsh diagram which may be constructed using these results. There are energy changes associated with some of these seven orbitals which are easy to understand in terms of overlap changes, but the most dramatic change is the one shown and is associated with the $3\sigma_g^+(s)$ orbital. It is strongly stabilized on bending as a result of mixing with a metal p orbital. One could visualize the stabilization on bending of a linear *sixteen* electron molecule as arising through a second order Jahn-Teller effect. The HOMO ($3\sigma_g^+$) is of symmetry σ_g^+ and the LUMO of π_u symmetry, so that a vibrational motion of π_u symmetry will couple the two. After bending there are now a total of eight orbitals to be filled leading to a sixteen electron count.

5.27. The construction of a molecular orbital diagram for this molecule is similar in principle to that used for $(AuPR_3)_6C^{2+}$ in Question 5.12. We use the analogy between the $AuPR_3$ unit and a hydrogen atom to construct the molecular orbital diagram for the trigonal bipyramidal $'H_5'$ (or $(AuPR_3)_5$) unit as in **5.54.** There are two a_1' orbital combinations, one where the axial and equatorial orbital sets are mixed in-phase and the other where they are mixed out-of-phase. Addition of the orbitals of the central nitrogen atom leads to the complete orbital picture as shown. Four bonding orbitals result which

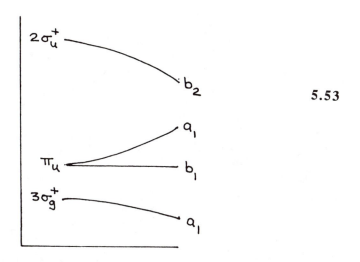

5.53

nicely accommodate the eight valence electrons of the molecule. The out-of-phase a_1' orbital combination remains roughly unchanged in energy, since it is approximately nonbonding with respect to the nitrogen $2s$ orbital. As in Li_4Me_4, question 5.15 and $(AuPR_3)_6C^{2+}$, question 5.12, there is probably little direct Au-Au bonding. The molecular orbital diagram gives some hints as to what favors the CH_5^+ structure over the trigonal bipyramidal one studied here. If the *HOMO-LUMO* gap of the trigonal bipyramid (**5.54**) is small then a second order Jahn-Teller distortion might be favored which couples the *HOMO* (e') with the *LUMO* (a_1'). The distortion mode determined by symmetry is of species $e' \times a_1' = e'$. Such a geometry change sends the $(AuPR_3)_5N^+$ structure to one related to that of CH_5^+ in the sense that two atoms of the polyhedron are brought close together. The driving force for the distortion will be set by the size of the *HOMO-LUMO* gap. This will increase as the electronegativity difference between the central atom and ligand increases. For C/H this might be expected to be small leading to a large driving force for distortion.

5.28 The conformational preferences for this species are readily obtained by consideration of the molecular orbital diagram of a Cp_2ML_2 molecule containing a 'bent' Cp_2M moiety. (The assembly of this diagram may be found in several places, including reference 1, page 396.) Basically the ligand, L, and CH_2 σ orbitals interact with and destabilize the b_2 and $2a_1$ fragment orbitals on Cp_2W (shown in **5.55**). This leaves the $1a_1$ fragment orbital to interact with and be stabilized by the p atomic orbital on the carbene in a π fashion. **5.56** shows the important effect of the location and shape of the filled $1a_1$ orbital of this molecule in determining the magnitude of this interaction. It is easy to see

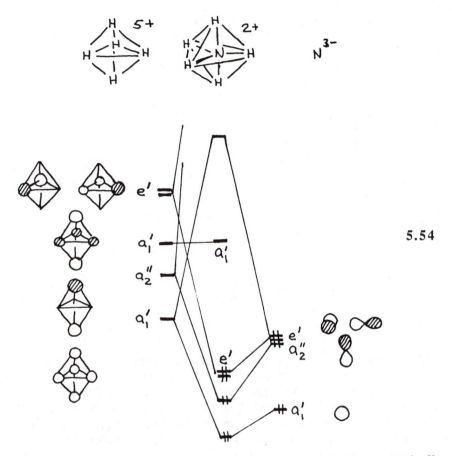

5.54

that overlap with the π orbital on CR_2 is maximized in the sterically unfavorable geometry.

5.29. (a) Electron counting gives 6 (Cr) + 8 $(4xCO)$ + 2 $(2x$ *one-electron carbon donors)* = 16, two short for a stable non square planar complex. Such metallacyclobutanes are in fact known but with 18 electrons, *e.g.*, $Ir(PMe_3)_3Br(CR'CR''CH_2)$. (b) **5.57** shows the construction of a molecular orbital diagram for the metallacyclobutane. There are two strong σ interactions which lead effectively to removal of two electrons from the set of $d\pi$ orbitals leading to a small HOMO-LUMO gap. (c) Distortion of the

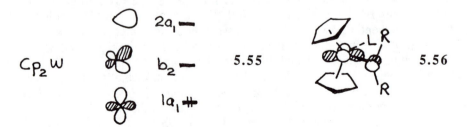

5.55

5.56

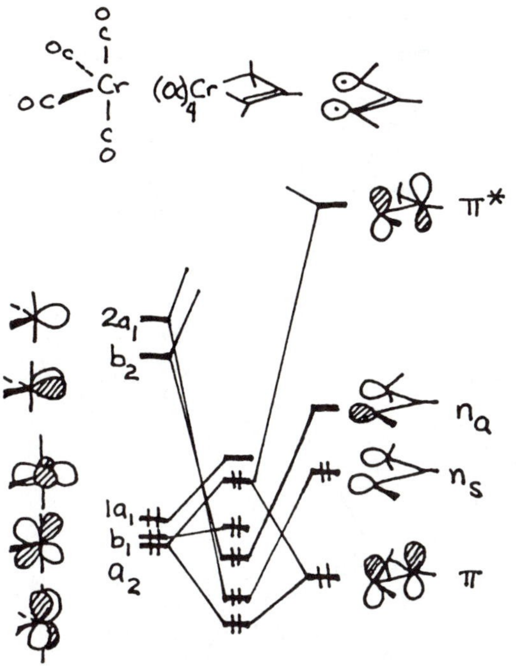

5.57

structure to the one shown in **5.16** leads to an electronic situation where the two one-electron carbon donors coordinated to the metal in **5.15** have been replaced by two two-electron donors, an olefinic double bond plus a carbene. Electron counting gives 6 (Cr) + 8 $(4xCO)$ + 4 $(2x$ *two-electron carbon donors)* = 18. A molecular orbital diagram connecting **5.15** and **5.16** is given in *New J. Chem.*, **15**, 769 (1991).

5.58

5.30.   A R_2P Lewis base would be regarded as in **5.58**. Thus the electron count around each Rh in the absence of any Rh-Rh bonds is: $2CO$ = 4e⁻, PR_2^- = 4e⁻, $Rh(1+)$ = d^8 = 8e⁻. Total = 16e⁻. In **5.17a** the local coordination geometry around each Rh is square planar, thus, with a 16e⁻ count there should be no Rh-Rh bond. In **b** one Rh is square planar, the other is *tetrahedral*. Thus, to make an 18e⁻ count at the tetrahedral Rh one must use 2 nonbonding electrons from the square planar metal to form a Rh-Rh single bond. The decrease of 1Å on going from **a** to **b** is consistent with this. Finally in **c** both Rh atoms are tetrahedral. Thus, we now need to share 2e⁻ from each Rh to make an 18e⁻ count and therefore a Rh-Rh double bond is expected. In agreement with this the Rh-Rh distance on going from **b** to **c** decreases by 0.21Å.

5.31   **5.59** shows the orbital interaction diagram for butadiene with $Fe(CO)_3$. There are two critical interactions; $2e_s$ with π_3 and $2e_a$ with π_2. In both cases

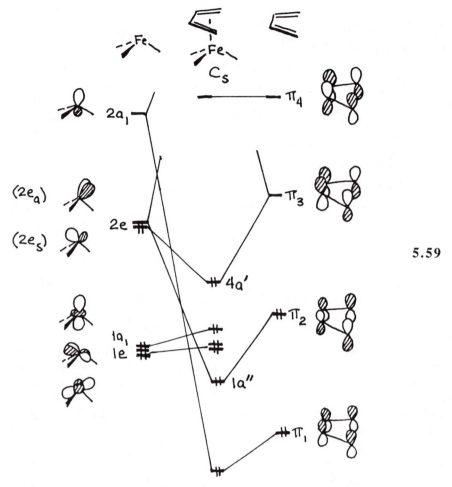

5.59

only the bonding combination is filled (*1a"* and *4a'*). Notice that π_2 is antibonding between the central carbon atoms in the butadiene ligand. Since π_2 is occupied, electron density is removed from π_2 in *1a"* which leads to an increase in the bond order between the central pair of carbon atoms. Likewise, since π_3 is empty, electron density flows from $2e_s$ to π_3 which is bonding between the central carbon pair. Hence the difference between the inner and outer carbon-carbon distances tends to be reduced when the ligand is complexed to $Fe(CO)_3$.

Chapter VI.

Organic Chemistry

6.1. Derive the symmetry species of the lowest-lying electronic states appropriate for square planar cyclobutadiene. Is the molecule expected to remain in a square geometry? Is the rule that asymmetric occupation of degenerate orbitals leads to geometrical instability (the Jahn-Teller theorem) violated here?

6.2 Determine the symmetry species of the irreducible representations obtained using the $p\pi$ orbitals of benzene as a basis. Construct symmetry adapted linear combinations of these orbitals, and, from their nodal properties rank them energetically. Add the correct number of electrons and see how the Hückel $4n + 2$ rule applies to the six-membered ring.

6.3 The CH_4^+ molecule can be made by ionization of the parent CH_4. We know that the neutral molecule is tetrahedral, but what is the geometry of CH_4^+? Derive the symmetry species of the ground electronic states of the molecule and its positive ion in the tetrahedral geometry and test them for first and second order Jahn-Teller instabilities. (The symmetry species of the bending mode which takes tetrahedral to square planar is e.)

6.4 By assembling a molecular orbital diagram from the two obvious fragments, show how in both conformations of the ethyl cation $(CH_3CH_2^+)$ there is a stabilization (hyperconjugation) via interaction of a CH bonding orbital on CH_3 with the empty p orbital on CH_2^+. Extend your answer to predict the more stable conformer for $CH_2FCH_2^+$ and $(Me_3Si)CH_2CH_2^+$.

*6.5 What do Jahn-Teller ideas have to say about the structures of benzene and its positive ion. Put aside your years of organic chemistry propaganda to get an objective approach for the former!

6.6. How might you stabilize a square planar CX_4 molecule, *i.e.*, what should X be?

6.7. By using first and second order perturbation theory and the π manifold of the molecule, predict which isomer, *cis* or *trans*, will be more stable for the substituted cyclobutadienes $X_2Y_2C_4$. Here X and Y are electron donating and withdrawing substituents.

6.8.   Construct a molecular orbital diagram for the π manifold of butadiene. Show why the inner C-C distance is longer than those for the outer ones and predict what will happen to the C-C distances upon addition or removal of electrons.

*6.9.   The $-CH_2SiMe_3$ ligand (X) when it replaces hydrogen in molecules with π systems, such as olefins or benzene, or lone pairs such as amines or sulfides, usually leads to quite a dramatic reduction in the corresponding ionization potential. For example, the shift from C_2H_4 to C_2H_3X is from 10.51 to 9.10 eV, and from C_6H_6 to $1,4$-$C_6H_4X_2$ it is from 9.24 to 7.75eV. The effect is usually ascribed to π/σ_{CSi} hyperconjugation, namely the destabilizing effect on the C-$C\pi$ bond shown in **6.1**. Notice that this leads to the opposite effect on the

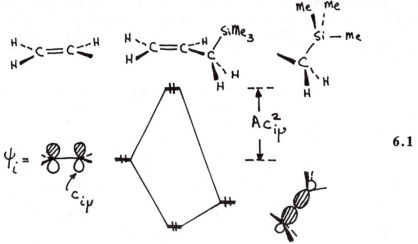

6.1

C-C π^* orbitals found for the SiR_3 group in Question 6.20.

Table 6.1   Substituted Benzene Ionization Patterns (eV)

	C_6H_6 1	1,2	1,3	1,4	1,3,5	1,2,4,5		1,2,3,4,5,6
IP_1	9.24	8.35	8.05	8.05	7.75	7.85	7.10	7.40
IP_2	9.24	9.00	8.55	8.40	8.75	7.85	7.75	7.40

Table 6.1 shows the first two ionization potentials for the substituted benzenes $C_6H_{6-n}X_n$. Perturbation theory tells us that the energy shift ΔE in orbital i will be proportional to the square of the coefficient $c_{i\mu}$ which describes the weight of the carbon $p\pi$ orbital at site μ in the orbital i. Show by means of a simple regression that $\Delta = IP(C_6H_{6-n}X_n) - IP(C_6H_6)$ is approximately given by an expression of the form $\Delta_i = -A\Sigma_\mu c_{i\mu}^2$. Show that a better fit is found if an inductive contribution is included, linear in the number of substituents, n; i.e., $\Delta_i = -A'\Sigma_\mu c_{i\mu}^2 + nB$. Determine numerical values for A and B.

6.10.   Using what you know about the structure of H_3^+ (Question 2.13 for example) predict a structure for the CH_5^+ molecule. This species has long been identified in the mass spectrometer but no experimental structural characterization has yet been made.

*6.11 (a) Use the molecular orbital diagram for benzene to determine the symmetry species and spin of the lowest energy electronic state for the ground electronic configuration. (b) Do the same for the excited state configurations arising from the excitation of a single electron. (c) Repeat (a) for the benzene dication, assuming it has the same structure as the neutral molecule (in fact it doesn't). (d) Predict the form of the electronic spectrum of benzene vapor. (e) Now imagine a crystal of benzene consisting of parallel layers of flat benzene molecules arranged as in a collection of pennies on a table. (It actually has a slightly more complex structure than this.) How would the electronic spectrum of benzene look recorded using light plane polarized parallel and perpendicular to the benzene planes?

6.12. The simplest argument, based on extrapolation of the series $H_3C\text{-}CH_3$, $H_2C=CH_2$, $HC\equiv CH$, is that there will be a quadruple bond between the carbon atoms in the C_2 molecule. Explore the orbital situation for diatomic molecules and identify the electronic conditions for this to be so. (C_2 in fact does not have a quadruple bond.)

6.13. A substituted derivative of the tetraradical shown in **6.2** has recently been prepared and studied. (a) Construct a molecular orbital diagram for the four orbitals $\chi_1\text{-}\chi_4$ of the molecule. Assume that all $C\text{-}C\text{-}C$ angles are 109.5°

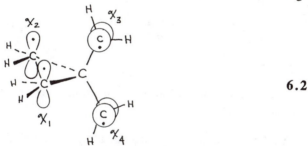

6.2

and all $H\text{-}C\text{-}H$ angles are 120°. With this geometry the values of $<\chi_1/\chi_3>$ etc., are quite large and those of $<\chi_1/\chi_2>$ etc., are smaller. (b) There are four electrons associated with the molecular orbitals in (a). Determine the symmetry of the candidates for the electronic ground state of the molecule. Does it have four unpaired electrons?

6.14. Construct a molecular orbital diagram for C_2R_4 by using the orbitals of the CR_2 fragments. Explore the electronic consequences of setting the $C\text{-}C$ π overlap equal to zero. A_2R_4 molecules are known for all Group 14 elements.

$$R\text{-}\!\!\diagup\!\!\diagdown A\text{--}A\diagdown\!\!\diagup R \longrightarrow$$

6.3

As the group is descended the geometry changes from the typical planar ethylene structure to that shown in **6.3** for Sn_2R_2. Show how this follows naturally from a decrease in the energetic importance of π bonding relative to σ bonding.

6.15. $CH_4{}^{2+}$ is a known molecule but its structure has not been experimentally determined. By a large *ab initio* calculation the geometry has been found to be that in **6.4a** which has a C_{2v} geometry. Another C_{2v} structure, **b**, was found to lie 13 kcal/mol higher in energy. The calculated geometric parameters are given in Table 6.2. Show why **a** is more stable than **b** and discuss the changes in bond lengths and angles on going from **a** to **b**. (Recall that the *H-H* distance in H_2 is 0.74Å.)

Table 6.2.

	a	b
C-H_1	1.28Å	1.43Å
C-H_2	1.14Å	1.12Å
H_1-H_4	1.03Å	0.90Å
H_1-C-H_4	47.7°	36.4°
H_2-C-H_3	124.1°	135.3°

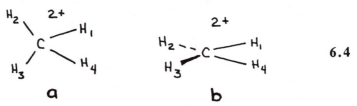

6.16. Using simple molecular orbital ideas calculate in kj/mol or kcal/mol the *C-H* bond energy in CH_4.

6.17. Construct molecular orbital diagrams for the π systems of both planar and twisted geometries for the allene $(CH_2 \cdot C \cdot CH_2)$ molecule. Show that a planar allene can be thought of as being Jahn-Teller unstable. Hence provide an orbital rationale for van't Hoff's rule. Decide on the lowest energy geometry for 2, 4, 6 and 8 π electrons.

*6.18. The photoelectron spectra observed for the π levels of a series of substituted benzenes are shown in **6.5** plotted as a function of the p orbital ionization energy, IP_x, of the substituent taken from atomic spectral data. Rationalize the qualitative form of this plot, and use first order perturbation theory to predict the slopes of the three lines.

6.19‡. Use a program which will generate the eigenvalues and eigenvectors of a real symmetric matrix to construct Hückel molecular orbital diagrams for the two isomers of borazine, $B_3N_3H_6$, shown in **6.6**. (a) Assume that the Hückel β values are equal for *B-B*, *N-N* and *B-N* contacts and equal to 5.0eV, and decide

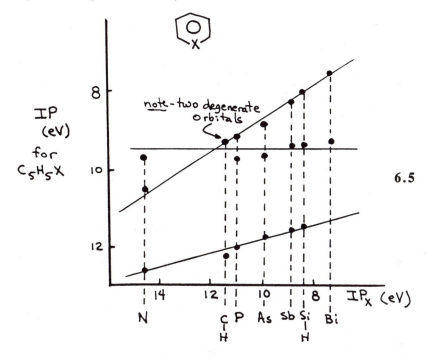

on the lower energy isomer. Devize a qualitative model to explain your numerical results. (b) Assume that the B-B, N-N and B-N interactions are different, by 10% at the most. Does this affect your answer? Ionization energies for atomic orbitals are given in Chapter I.

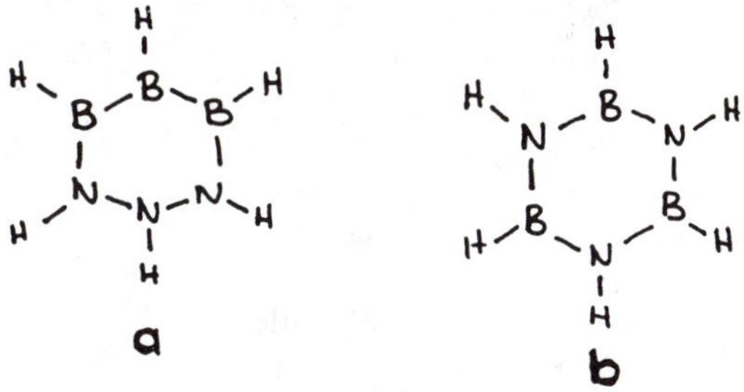

6.6

6.20. **6.7** shows the epr spectra of two related radical anions of substituted benzenes. The fine structure shown in (a) has been interpreted as arising via coupling to the CH_3 groups attached to the silicon atoms. This indicates that the orbital containing the unpaired electron has a large orbital contribution from silicon, and a much smaller one from the ring carbon atoms. The spectrum in (b) may be interpreted using the opposite view. Here there is no methyl contribution at all; the electron density resides completely on the ring carbon atoms. Recognizing that the electronegativity of carbon is higher than that of silicon, rationalize these results. What does your model predict for the products of Birch reduction of the two molecules.

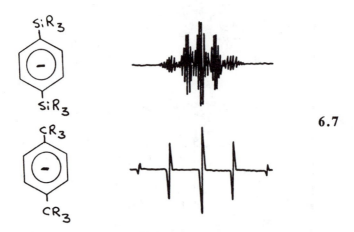

6.7

6.21. **6.8** shows the structure of a substituted ethylene dication, $[C(N(CH_3)_2)_2]_2^{+2}$. Why is the N_4C_2 skeleton twisted, and why is the central C-

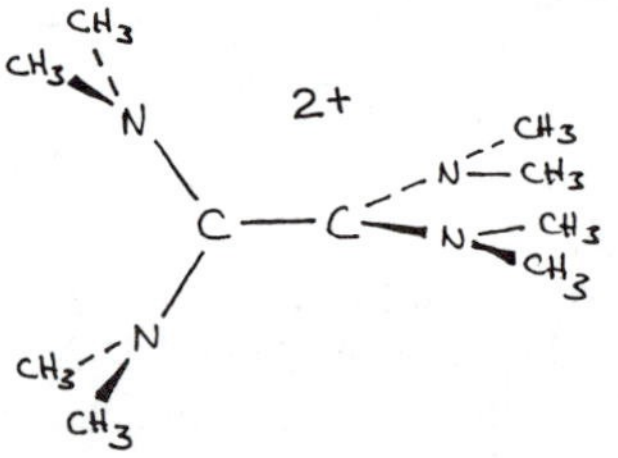

6.8

C distance much longer (1.55Å) than in the neutral molecule (1.36Å)? Notice that the $(CH_3)_2N$ groups are planar. Use a three center orbital model to show why the CN distance shortens dramatically (1.47Å to 1.36Å) on formation of the dication. Is the result of your calculation in accord with the numerically computed one that on formation of the dication the largest change in atomic charge is associated with the central carbon atoms?

6.22. The bismethylenecyclobutadiene molecule, C_6H_6, shown in **6.9**, has recently been synthethized. It is interesting in that while there are six π electrons in six molecular orbitals, no resonance structure can be drawn with three double bonds. (a) Determine the relative energies and orbital shapes for the six π orbitals, given that all the C-C distances are equal. (b) Show the orbital occupations and determine the symmetries of what you think are the

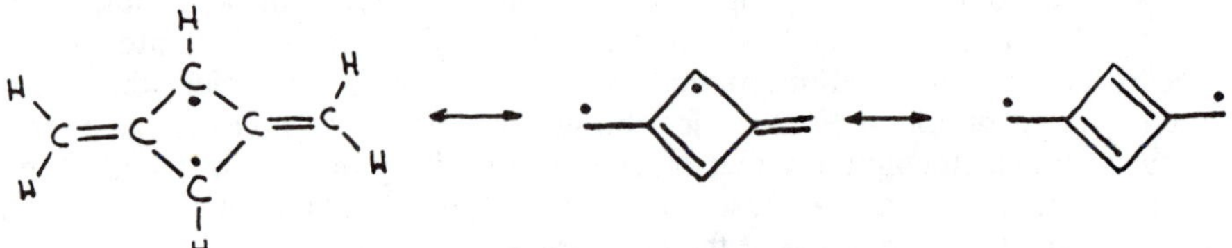

6.9

lowest singlet and triplet electronic states for the molecule.

6.23. $C_5H_5^-$ is a planar molecule with three coordinate carbon atoms but $C_5H_5^+$ has a cage structure which contains one five coordinate carbon atom. Explain.

6.24. In cyclohexanol the O-H group is equatorial but in 2-hydroxytetrahydropyran it is axial. By considering the interaction of the empty σ-antibonding orbital of C-OH with a filled ring oxygen lone pair orbital, show how this (anomeric effect) comes about.

Answers

6.1. This is perhaps the quintessential organic chemistry Jahn-Teller problem, but is more complex than it appears at first sight. The π molecular orbital diagram for the molecule is shown in **6.10**. The π electronic

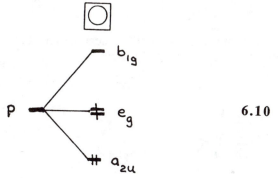

6.10

configuration is just $(a_{2u})^2(e_g)^2$. Determination of the symmetry species of the singlet and triplet states of the molecule requires evaluation of the symmetric and antisymmetric direct products appropriate for $e_g \times e_g$. This is done in Table 6.3 and leads directly to the generation of a total of four electronic states,

Table 6.3

for e_u	E	2C_4	C_2	2C_2'	2C_2"	i	2S_4	σ_h	2σ_v	2σ_d	
$\chi(\mathcal{R})$	2	0	-2	0	0	-2	0	2	0	0	
$\chi(\mathcal{R}^2)$	2	-2	2	2	2	2	-2	2	2	2	
$\chi^2(\mathcal{R})$	4	0	4	0	0	4	0	4	0	0	
$\chi^2_{sym}(\mathcal{R})$	3	-1	3	1	1	3	-1	3	1	1	$a_{1g} + b_{1g} + b_{2g}$
$\chi^2_{antisym}(\mathcal{R})$	1	1	1	-1	-1	1	1	1	-1	-1	a_{2g}

three singlets, $^1A_{1g}$, $^1B_{1g}$, $^1B_{2g}$, and a triplet $^3A_{2g}$. First we notice that the triplet state is not orbitally degenerate and thus is geometrically stable on first order Jahn-Teller grounds. However, notice too that *no* degenerate electronic singlet state is produced either. Thus the first order Jahn-Teller theorem is not applicable at all to this problem. However the three singlet states may well be close in energy; they differ by two-electron terms in the Hamilitonian only, namely the ones which involve the electron-electron interactions, $1/r_{ij}$. A second order Jahn-Teller interaction arising *via* the term $|<X|(\partial \mathcal{H}/\partial q)_0|Y>|^2/ (E_X^{(0)} - E_Y^{(0)})$ is possible. Since this second order interaction couples electronic states arising from the same electronic configuration we call it a *pseudo* Jahn-Teller effect. This will result in a stabilization of the electronic ground state, X, as a result of mixing with an excited state, Y, *via* a distortion, q, if the symmetry of the integrand in the numerator is of a_1 symmetry. This will only be the case if the direct product $\Gamma_X \times \Gamma_Y$ contains Γ_q. But what is the symmetry species of the lowest singlet state? This is difficult to predict without a calculation but if either the $^1A_{1g}$ or $^1B_{1g}$ state lies lowest then there will be a distortion of b_{2g} symmetry, shown in **6.11** which will take the square to the

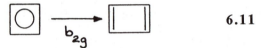

6.11

rectangle. Thus the rule that asymmetric occupation of degenerate orbitals leads to a geometrical distortion is not violated.

6.2.   **6.12** shows the basis orbitals we will use with the D_{6h} point group. They

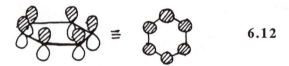

6.12

transform as:

D_{6h}	E	$2C_6$	$2C_3$	C_2	$3C_2'$	$3C_2''$	i	$2S_3$	$2S_6$	σ_h	$3\sigma_d$	$3\sigma_v$
	6	0	0	0	-2	0	0	0	0	-6	0	2

which leads to the irreducible representations $a_{2u} + e_{1g} + e_{2u} + b_{2g}$. The generation of the wavefunctions is described in many places (e.g., reference 5). They are, neglecting overlap in the normalization:

$$\psi(a_{2u}) = 6^{-1/2}(\phi_1 + \phi_2 + \phi_3 + \phi_4 + \phi_5 + \phi_6)$$
$$\psi(b_{2g}) = 6^{-1/2}(\phi_1 - \phi_2 + \phi_3 - \phi_4 + \phi_5 - \phi_6)$$
$$\psi(e_{1g}) = 12^{-1/2}(2\phi_1 + \phi_2 - \phi_3 - 2\phi_4 - \phi_5 + \phi_6)$$
$$\psi(e_{1g})' = 2^{-1}(\phi_2 + \phi_3 - \phi_5 - \phi_6)$$
$$\psi(e_{2u}) = 12^{-1/2}(2\phi_1 - \phi_2 - \phi_3 + 2\phi_4 - \phi_5 - \phi_6)$$
$$\psi(e_{2u})' = 2^{-1}(\phi_2 - \phi_3 + \phi_5 - \phi_6)$$

and are sketched out in **6.13** in energetic order. As the number of nodes increases so does the energy. Notice that the e_{1g} pair is bonding overall and the e_{2u} pair antibonding. Thus the six electrons form three bonding pairs and

b_{2g} —

e_{2u} =

e_{2g}

a_{2u}

6.13

the Hückel rule thus represents filling of all the bonding orbitals.

6.3. **2.30** shows a molecular orbital diagram for the tetrahedral geometry. The electronic configuration of the neutral molecule is $(1a_1)^2(1t_2)^6$, leading to a $^1A_1$ electronic state. Since this is non-degenerate a first order Jahn-Teller distortion is inapplicable. Since the *HOMO-LUMO* gap is large a second order Jahn-Teller distortion is not energetically important. The electronic configuration of the cation is $(1a_1)^2(1t_2)^5$. The symmetry species of the electronic state is thus $^2T_2$ and so a first order Jahn-Teller effect is predicted. The symmetry species of the distortion will be given *via* the symmetric direct product $t_2 \times t_2$. shown in Table 6.4.

Table 6.4.

for t_2	E	$8C_3$	$3C_2$	$6S_4$	$6\sigma_d$
$\chi(\mathcal{R})$	3	0	-1	-1	1
$\chi(\mathcal{R}^2)$	3	0	3	-1	3
$\chi^2(\mathcal{R})$	9	0	1	1	1
$\chi^2_{sym}(\mathcal{R})$	6	0	2	0	2

This is simply equal to $a_1 + e + t_2$, and thus modes of $e + t_2$ symmetry may relieve the instability. The degeneracy is thus relieved during an e species motion of the form required to take the tetrahedron towards the square plane. The Jahn-Teller theorem only tells us about the symmetry of the distortion, but not its direction or its magnitude. Likewise, it does not tell us whether the distortion to large *H-A-H* angles will stop at an intermediate D_{2d} or proceed all the way to a D_{4h} geometry. In fact the experimental evidence is for an small distortion towards the D_{2d} arrangement. A t_2 motion can lead to yet another geometry, that with C_{2v} symmetry.

6.14

6.4   The deepest lying orbitals of the two fragments are shown in **6.14**, those of CH_3 at the left and those of the two conformations of CH_2^+ at the right (a and b). The antibonding orbitals, lying high in energy, are not shown, since their energy differences relative to the bonding set are so large that the interaction with them will be weak. The interactions between the cylindrically symmetrical orbitals $\sigma(CH_3)$, $\sigma(CH_2)$ and n_σ, will play no part in determining the conformational preferences. It is the interactions which develop between the $\pi(CH_3)$ orbitals on CH_3 and the $\pi(CH_2)$ and n_p orbitals of CH_2^+ which will determine the stereochemistry. In conformation a, the carbon p_y component of the $\pi(CH_2)$ orbital and the carbon p_x orbital which makes up $n_p{}^x$ can interact with symmetry appropriate orbitals on CH_3, namely $\pi(CH_3{}^y)$ and $\pi(CH_3{}^x)$ respectively. The result is a two-electron stabilizing interaction and one four-electron destabilization (**6.15**). The first of the two is the hyperconjugative interaction, shown in **6.16**. The analogous result for conformation b, where the CH_2^+ group is rotated around the C-C axis by 90° is also shown. All that has changed are the x and y labels of the CH_2^+ orbitals, but each of them finds a symmetry match with the pair of $\pi(CH_3)$ orbitals on CH_3. The picture is thus identical for both conformations resulting from the degeneracy of the $\pi(CH_3)$ orbitals on CH_3. The hyperconguative interaction is identical for the two conformers and leads to free rotation of the CH_2 unit around the C-C axis in the ethyl cation.

     The situation is different when the CH_3 group is replaced by CH_2R. If the substituent is such that the $\pi(CH_3{}^x)$ orbital is pushed to lower energy then the hyperconjugative interaction is reduced in conformation a and conformation b is favored. This is the case for $R = F$. If the $\pi(CH_3{}^x)$ orbital is

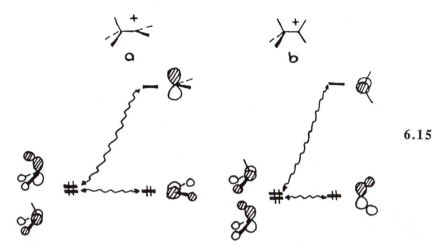

6.15

pushed to higher energy then the hyperconjugative interaction is increased in conformation **a** and this conformation is favored. This is the case for $R = Me_3Si$.

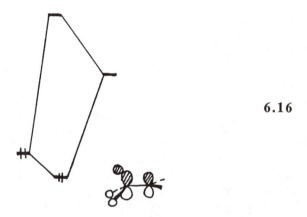

6.16

6.5. The π levels of benzene are shown in **6.13**. The ground electronic state is $^1A_{1g}$, which is clearly stable on first order Jahn-Teller grounds. However if the first excited state, arising *via* promotion of an electron from e_{1g} to e_{2u} is not too high in energy then a second order Jahn-Teller effect is possible. Let's see what the symmetry species of the distortion coordinate q needs to be such that the term $|<X|(\partial \mathcal{H}/\partial q)_0|Y>|^2|(E_Y^{(0)} - E_X^{(0)})$ is non-zero. All we need is evaluation of the direct product $e_{1g} \times e_{2u}$. This is easy to do and the result is $b_{1u} + b_{2u} + e_{2u}$. These are then the possible motions which will lead to a second order Jahn-Teller stabilization. **6.17** shows motions of these symmetries. (From reference 3 p93.) Especially interesting is the one of b_{2u} symmetry which shortens and lengthens alternate bonds, and which will eventually lead to three acetylene molecules. We know however that benzene is perfectly stable in its regular hexagonal geometry, and so for this particular case the second order Jahn-Teller effect does not lead to a geometry change.

6.17

However, benzene may be an exception. Exactly the same arguments apply to the isoelectronic molecule N_6 and indeed to the cyclic H_6 molecule (see Question 2.4). Neither of these exist, but the products of the distortion, three N_2 or H_2 molecules, are of course the normal forms of these molecules. This leads to some comments concerning Jahn-Teller distortions in general. Consideration of the symmetry and occupation of the *HOMO* and the symmetry species of the *LUMO* are important in the identification of likely distortion routes. Very often such considerations involve sets of orbitals which are delocalized; the π orbitals in benzene for example. However, there are other deeper-lying occupied orbitals which are often important too. Such occupied sets of localized orbitals (the *C-C* σ orbitals of benzene, for example) invariably resist such asymmetric distortions. If we regard each localized bond as being described by a harmonic potential, $V = 1/2kx^2$, where x is the displacement from equilibrium, then the 'elastic' energy of benzene with its six σ bonds is equal to $3kx^2$. Alternately stretching and contracting the bonds around the ring by Δx, leads to a new energy of $3k(x^2 + \Delta x^2)$. This is minimized for $\Delta x = 0$. Thus, contrary to the 'folklore' of organic chemistry, benzene is held in its regular geometry by the energetic preferences of the σ manifold rather than those of the 'resonating' π network. (For a detailed discussion of this problem see *J. Amer. Chem. Soc.*, **107**, 3089, (1985), *J. Org. Chem.*, **51**, 3908 (1986) and *J. Mol. Structure (Theochem.)*, **229**, 197 (1991).)

The benzene positive ion, with three electrons in the e_{1g} orbital has a $^2E_{1g}$ electronic ground state. This is Jahn-Teller unstable. The symmetry of the possible distortion modes are given by the symmetric direct product of $e_{1g} \times e_{1g}$, evaluated in Table 6.6 as $a_{1g} + e_{2g}$. Vibrations with these symmetries are shown in **6.17**. Just as for the benzene dication in Question 6.11, the molecule should distort away from the symmetrical structure.

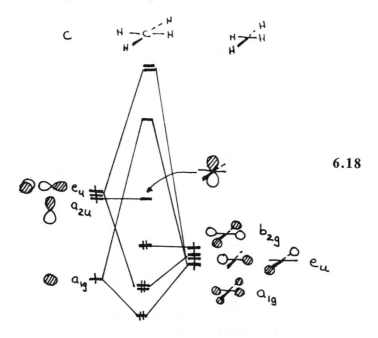

6.18

6.6. Here we need to make use of the molecular orbital diagram for square planar CH_4 derived in **6.18**. Clearly the molecule is energetically unfavorable because of the presence of the nonbonding b_{1g} orbital and an uninvolved high-lying a_{2u} orbital. The first strategy is to populate the a_{2u} rather than the b_{1g} orbital by making the ligands less electronegative. (In this way a_{2u} lies lower than b_{1g}.) The presence of low-lying acceptor orbitals on the atoms attached to carbon will then stabilize this orbital, perhaps leading to a stable square planar geometry. **6.19** shows how this occurs. It is just the same mechanism used to stabilize a planar NR_3 unit in Question 4.11. In fact calculations suggest that the molecule trilithiomethane, CLi_3H, will be planar.

6.7. **6.20** shows the π orbitals of cyclobutadiene and how they change in energy on substitution by ligands which effectively change the Hückel α value of the adjacent carbon $p\pi$ orbital. We have an infinite set of choices for the doubly degenerate e_g set. Two different but equally useful orbital pairs are shown, one appropriate for each substitution pattern. The first order energy shifts are are shown in terms of $\delta\alpha$, the magnitude of the shift in the Hückel α on attachment of X and Y. (For simplicity we assume that they are equal and

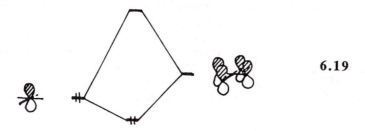

6.19

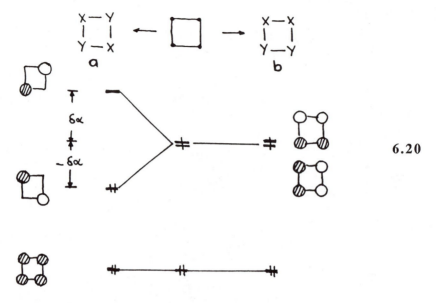

opposite, athough it makes no difference to our argument.) These first order shifts are just $\Sigma_i c_{ij}^2 \delta\alpha$ for each orbital j where the summation is over all of the atomic orbitals i. As can be seen they are very different for the two substitution patterns. For the e_g set using the *trans* substitution, **a**, the shift is just $\Sigma_i c_{ij}^2 \delta\alpha = 2(1/\sqrt{2})^2 \delta\alpha = \delta\alpha$ for one component and $2(1/\sqrt{2})^2(-\delta\alpha) = (-\delta\alpha)$ for the other. Using the *cis* substitution **b** the shift is $2(1/2)^2 \delta\alpha - 2(1/2)^2 \delta\alpha = 0$. Thus the pattern **a** is favored. Calculations (*Chem. Comm.*, 240 (1969)) suggest energy differences (*e.g.*, for the *B/N* compound) of 3eV.

6.8.   The assembly of the molecular orbital diagram for the π manifold of butadiene is very similar indeed to that for the linear H_4 molecule described in Reference 1, chapter 1, page 63. First generate an approximate level diagram by separately interacting the two σ orbitals of the two H_2 fragments with each other, followed by the two σ^* orbitals as in **6.21** to give two pairs of σ_u^+ and σ_g^+ orbitals. With four electrons and this molecular orbital diagram we would predict a bond order of zero for the central linkage since the bonding interaction in the deepest-lying orbital is cancelled exactly by the antibonding interaction in the *HOMO*. However this picture neglects interaction between the σ and σ^* orbitals on the two fragments. So now we allow in turn the two σ_g^+ orbitals to interact and the the two σ_u^+ orbitals to interact in second order. The lower one in each case is stabilized and the upper one destabilized. The new form of the molecular orbitals are simply obtained by mixing the upper into the lower in a stabilizing way in each case (**6.22**). The result is that the central bond is strengthened and the terminal ones weakened compared to the first case, but since this is a second order effect, the overall prediction of the relative strength of the two bonds predicted from **6.21** holds. Clearly addition of electrons will strengthen the middle bond and weaken the terminal ones, whereas removal of electrons will lead to the converse.

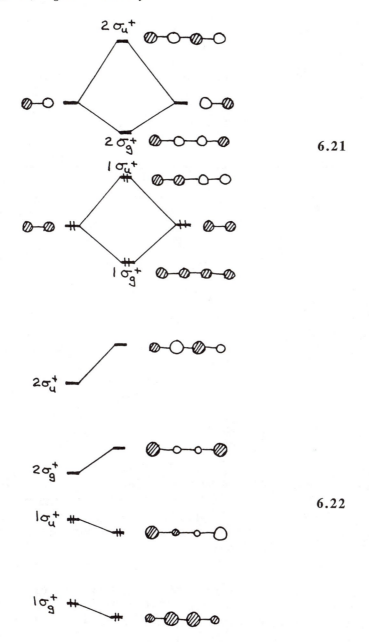

6.21

6.22

6.9.   The two π orbitals under consideration are shown in **6.23** along with the values of $c_{i\mu}^2$ at each site. Table 6.5 shows the values of $\Sigma_\mu c_{i\mu}^2$ and n for each species.

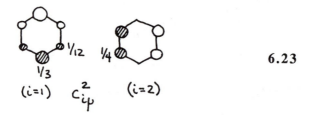

6.23

Table 6.5

	C_6H_6 1	1,2	1,3	1,4	1,3,5	1,2,4,5	1,2,3,4,5,6
Σ_1	0	1/3	1/2	1/2	2/3	1/2	1	1
Σ_2	0	0	1/6	1/6	0	1/2	1/3	1
n	0	1	2	2	2	3	4	6

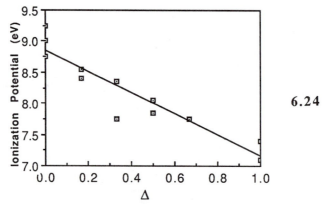

6.24

6.24 shows a plot of IP vs $\Delta_i = -\Sigma_\mu c_{i\mu}^2 A$ from which $A = 1.68eV$. A regression using $\Delta_i = -\Sigma_\mu c_{i\mu}^2 A + nB$ led (*Angew. Chem.*, **28**, 1627 (1989)) to values of $A = 1.33eV$ and $B = 0.24eV$. For methyl substituents the corresponding parameters are 0.86eV and 0.09eV. The large difference in B is due to the electronegativity difference between H and Si, and the smaller value for A in the methyl case due to the reduced importance of this hyperconjugative interaction.

6.10. H_3^+ is a molecule whose σ levels are identical in form and occupancy to the π levels of the cyclopropenium cation $C_3H_3^+$. The molecular orbital diagram is assembled in **6.25** from $H_2 + H^+$. Such a decomposition immediately suggests the substitution of an isolobal species for H^+. One such fragment is of course CH_3^+, leading to the geometry of the CH_5^+ ion of **6.26**. This is the lowest energy geometry from calculations.

6.11. (a) The ground electronic configuration of benzene is $(a_{2u})^2(e_{1g})^4$, which, as a closed shell gives rise to a $^1A_{1g}$ state. (b) The lowest excited states arise from the configuration $(a_{2u})^2(e_{1g})^3(e_{2u})^1$. The symmetry species of these states arise *via* the direct product $e_{1g} \times e_{2u}$, which is simply evaluated as $b_{1u} +$

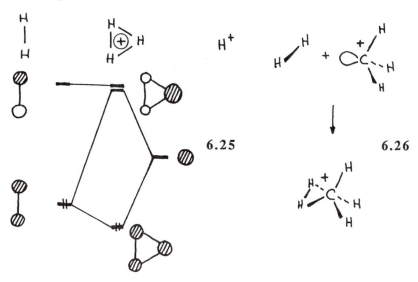

6.25

6.26

$b_{2u} + e_{1u}$, leading to the electronic states $^1B_{1u} + {}^1B_{2u} + {}^1E_{1u}$ and $^3B_{1u} + {}^3B_{2u}$ + $^3E_{1u}$. A set of singlets and triplets with the same spatial symmetry arise since there are no restrictions from the Pauli Principle concerning the way the levels are occupied with electrons. The triplets lie lower in energy because of their more favorable exchange energy. (b) The benzene dication has the electronic configuration $(a_{2u})^2(e_{1g})^2$. Now we need to use the symmetric and antisymmetric direct products to generate the symmetry species of the singlet and triplet electronic states since the Pauli Principle restricts the way the electrons may be placed in the orbitals. This is simple to do and is shown in Table 6.6 We just need to remember that;

$$\chi^2{}_{sym}(\mathcal{R}) = (\chi^2(\mathcal{R}) - \chi(\mathcal{R}^2))/2 \quad and \quad \chi^2{}_{antisym}(\mathcal{R}) = (\chi^2(\mathcal{R}) - \chi(\mathcal{R}^2))/2$$

Table 6.6

for e_{1g}	E	$2C_6$	$2C_3$	C_2	$2C'_2$	$3C''_2$	i	$2S_3$	$2S_6$	σ_h	$3\sigma_d$	$3\sigma_v$
$\chi(\mathcal{R})$	2	1	-1	-2	0	0	2	1	-1	-2	0	0
$\chi(\mathcal{R}^2)$	2	-1	-1	2	2	2	2	-1	-1	2	2	2
$\chi^2(\mathcal{R})$	4	1	1	4	0	0	4	1	1	4	0	0
$\chi^2{}_{sym}(\mathcal{R})$	3	0	0	3	1	1	3	0	0	3	1	1
$\chi^2{}_{antisym}(\mathcal{R})$	1	1	1	1	-1	-1	1	1	1	1	-1	-1

From inspection of the character table this leads to $^3A_{2g}$, $^1A_{1g}$ and $^1E_{2g}$ states. Thus we can say that since the the triplet state is not orbitally degenerate it will be Jahn-Teller stable. However the $^1E_{2g}$ state, probably the lowest singlet state, will be Jahn-Teller unstable. We leave it to the reader to show that the symmetric direct product $e_{1g} \times e_{1g} = a_{1g} + e_{2g}$. The latter is the symmetry species of the mode which will relieve the Jahn-Teller instability. The two components of the two e_{2g} vibrations associated with the C_6 skeleton are shown in **6.17**. The geometry of the molecule will certainly not be a regular hexagon.

(c) The intensity of the electronic absorptions of the neutral benzene molecule will be determined by the magnitude of the integrals $<X/\mu/Y>$ where X and Y are the ground and excited states respectively and μ is the dipole moment operator. The integral will be nonzero only if the spin multiplicity óf the two states are the same and if Γ_μ is contained in $\Gamma_X \times \Gamma_Y$. For D_{6h} $\mu \Rightarrow a_{2u} + e_{1u}$. Thus only the transition $^1A_{1g} \rightarrow {}^1E_{1u}$ will be allowed, and one strong band will be found in the electronic absorption spectrum.

(d) In the hypothetical solid we need to remember that the absorption intensity is proportional to $cos^2\theta$, where θ is the angle between the oscillating electric vector of the incoming radiation and the transition dipole moment of the molecule. Since $\mu_{x,y} \Rightarrow e_{1u}$, if the incoming radiation is polarized perpendicular to the plane of the benzene molecule ($\theta = 90°$) the intensity will be zero.

6.12.   The only way we can see that this might occur is if the interaction between the valence s orbitals is much larger than that between the $p\sigma$ orbitals and/or the s/p energy gap is small with respect to the interaction. This is shown in the orbital diagram of **6.27**. It is not very likely.

6.13. (a) The point symmetry of the molecule is D_{2d}. The basis orbitals transform as shown in Table 6.7.

Table 6.7

D_{2d}	E	$2S_4$	C_2	C_2'	σ_d	
Γ	4	0	0	0	-2	$a_2 + b_1 + e$

The properly symmetry adapted functions which result are shown in **6.28**. If we focus on the interactions with the largest overlaps, namely $<\chi_1/\chi_3>$ *etc.*, then a_2 is antibonding, b_1 is bonding and e is weakly bonding ($<\chi_1/\chi_2>$ is

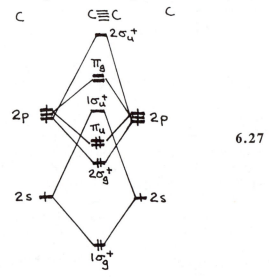

6.27

$$\psi_{a_2} \propto \mathcal{X}_1 - \mathcal{X}_2 + \mathcal{X}_3 - \mathcal{X}_4 \equiv$$

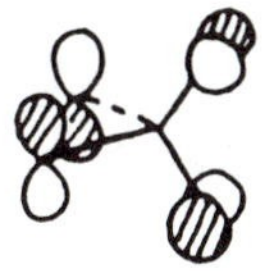

$$\psi_{b_1} \propto \mathcal{X}_1 - \mathcal{X}_2 - \mathcal{X}_3 + \mathcal{X}_4 \equiv$$

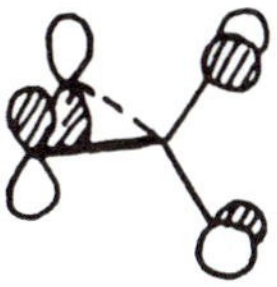

6.28

$$\psi_e \propto \mathcal{X}_1 + \mathcal{X}_2 \qquad \equiv$$

$$\psi_e' \propto \mathcal{X}_3 + \mathcal{X}_4 \qquad \equiv$$

small).

(b) Using the result from (a) the molecular orbital diagram may be constructed as in **6.29**. The only plausible ground state electronic

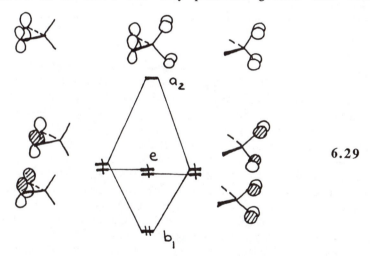

6.29

configuration is $(b_1)^2(e)^2$. The symmetry species of the possible electronic states are generated by constructing the symmetric and antisymmetric direct products of e as in Table 6.8.

This leads to the electronic states $^1A_1$, $^1B_1$, $^1B_2$ and $^3A_2$. If one of the singlet states lies lowest then it is possible that there will be a pseudo Jahn-Teller distortion which will lead to molecular distortion away from tetrahedral symmetry. With only one component of e occupied, we can see from **6.29** that a likely distortion is a scissors motion which will stabilize one component and destabilize the other (**6.30**). If the molecule is a triplet then a tetrahedral

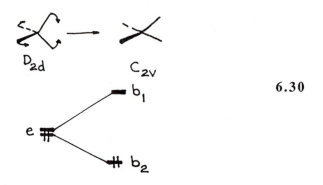

D_{2d} C_{2v}

6.30

Table 6.8

for e	E	$2S_4$	C_2	C_2'	σ_d	
$\chi(\mathcal{R})$	2	0	-2	0	0	
$\chi(\mathcal{R}^2)$	2	-2	2	2	2	
$\chi^2(\mathcal{R})$	4	0	4	0	0	
$\chi^2_{sym}(\mathcal{R})$	3	-1	3	1	1	$= a_1 + b_1 + b_2$
$\chi^2_{antisym}(\mathcal{R})$	1	1	1	-1	-1	$= a_2$

geometry is expected.

6.14. This orbital construction is particularly simple. Using the orbitals for CH_2 on the right side of **6.14**, the two n_σ orbitals combine to form σ and σ^* combinations, and the two n_p orbitals yield π and π^*. There are a total of four electrons here so that σ and π are filled. A large π overlap integral between the two fragment orbitals leads to a large $HOMO(b_{3u})$-$LUMO(b_{2g})$ gap. (In ethylene the $\pi \rightarrow \pi^*$ transition occurs in the vacuum ultraviolet.) As the π overlap decreases so this $HOMO$-$LUMO$ gap decreases. The second-order Jahn-Teller approach would predict a distortion of $b_{3u} \times b_{2g} = b_{1u}$ symmetry. As shown in **6.31** such a motion takes each planar center to a pyramidal one. **6.32** shows the overlap of the CH_2 fragment orbitals in this, and the related *trans*, geometries. As is quite clear from this picture, two 'bent' σ bonds have replaced the one σ and one π of the planar structure. The relative importance of π relative to σ bonding decreases on moving down a main group of elements, but the distortion of the molecule away from planar is aided by another effect too. The inversion barriers for second period (B to O) hydrides is considerably smaller than their third row (Al to S) analogues (see Question 4.11). For example, whereas CH_3 is planar, SiH_3 is pyramidal. Similarly, chirally pure simple amines cannot be isolated, but their anaolgous phosphines

b_{1u}

6. 31

6.32

may. Thus pyramidalization at each center on σ grounds is energetically encouraged on moving below the second period elements.

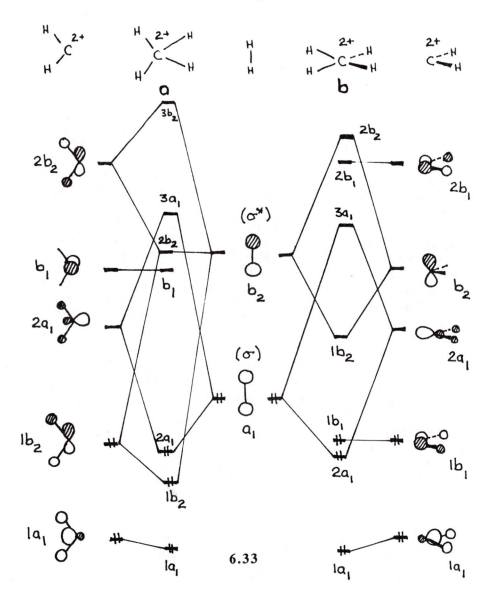

6.33

$\psi_{1b_2} \approx$ +   +   =

$\psi_{2b_2} \approx$ +   +   =   6.34

$\psi_{3b_2} \approx$ +   =

6.15. In both instances the H_1-H_4 distance is short. A natural way to look at the bonding in both molecules is to consider the interaction of CH_2^{2+} with H_2. This is done in **6.33** for the two geometries. In both structures H_2 σ interacts primarily with the $2a_1$ fragment on CH_2^{2+} to produce a stabilized bonding orbital (molecular $2a_1$) which is filled, and and empty antibonding combination (molecular $3a_1$). The overlap (and of course the energy gaps) between H_2 σ and $2a_1$ do not change on going from **a** to **b**. What does change is the way H_2 σ^* interacts with the CH_2^{2+} fragment. In **a** H_2 σ^* can interact with the $1b_2/2b_2$ CH_2^{2+} set to produce bonding $1b_2$, a nonbonding $2b_2$ and an antibonding $3b_2$ combinations (**6.34**). Only the molecular $1b_2$ bonding combination is filled. In **b** H_2 σ^* can now only interact with the nonbonding p atomic orbital on CH_2^{2+} (b_2) to produce bonding $1b_2$ and antibonding $2b_2$ combinations (**6.35**). Both molecular $1b_2$ and $2b_2$ are empty in **b**. (Note that it is not reasonable to presume in **b** that the $1b_2$ molecular orbital lies lower in energy than $1b_1$. The C-H_2 distance (1.12Å) is much shorter than the C-H_1 (1.43Å) distance. Thus, while both $1b_1$ and $1b_2$ are both C-H σ bonding molecular orbitals, the former must lie at a lower energy because of the shorter C-H_1 distance. In **a** the CH_2^{2+} σ_π orbital is stabilized by H_2 σ^* whereas in **b** it is not. Therefore, **a** must be more stable than **b**. The size of the *HOMO-LUMO* gaps produced also suggests stbility of **a** over **b**. In both molecules there will be significant H_2 $\sigma/2a_1$ CH_2^{2+} interaction. Thus electron density flows (**6.36**) from the H_2 unit towards CH_2^{2+} which leads to a weakening of the H_1-H_4 bond. As a result the H_1-H_4 distance is longer than in the free H_2 molecule. The H_1-C-H_4 angle is larger and the H_1-H_4 distance longer in **a** than in **b** because H_2 σ^* is partially occupied as

$\psi_{1b_2} \approx$ +   =

$\psi_{2b_2} \approx$ +   =   6.35

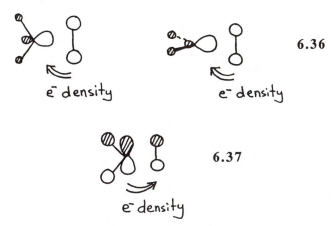

6.36

e⁻ density   e⁻ density

6.37

e⁻ density

in **6.37** in **a** but not in **b**. This extra interaction also makes the C-H_1 distance shorter in **a**. In this picture electron density flows from a C-H bonding orbital on CH_2^{2+} to H_2. Thus the C-H_2 distance is slightly longer in **a**. Secondly this interaction is facilitated if the $1b_2$ fragment orbital in **a** is raised in energy since the gap between $1b_2$ and H_2 σ^* becomes smaller. One way to do this is to make the H_2-C-H_3 angle smaller as in **6.38**. Therefore in **a** where 'backdonation' can be significant, the H_2-C-H_3 angle is smaller than in **b** where 'backdonation' is not.

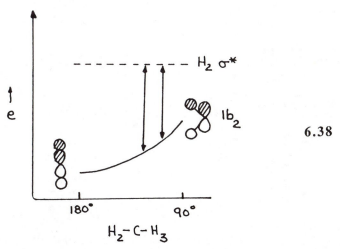

6.38

6.16. At present such an estimate lies beyond both one-electron and *SCF* calculations which do not include configuration interaction. A quantitative understanding of bond energies, which would eventually lead to enormous advances in the study of reaction kinetics and mechanisms is one of the greatest challenges facing the theoretical chemist today.

6.17. Orbital diagrams for these two geometries are shown in **6.39**. The π orbitals here are strictly analogous to the σ orbitals of H_2 and linear H_3. Notice that $1b_{1u}$ in D_{2h} is lower in energy than the $1e$ set is in D_{2d}, and $2b_{1u}$ is much higher in energy than the $2e$ set. (Hückel theory gives the energy levels

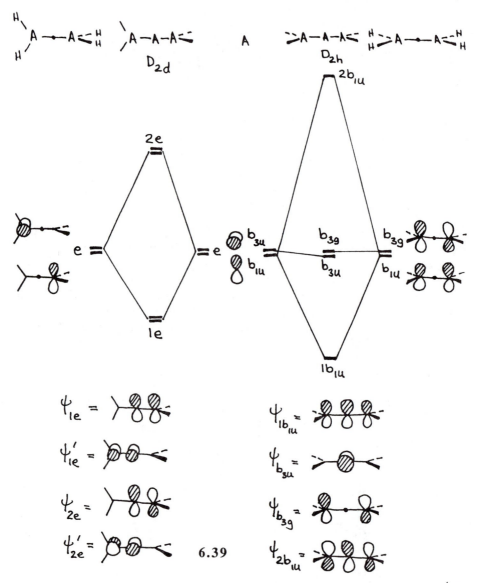

6.39

as $\alpha \pm \beta$ (twice) for the twisted form (two ethylene units) and α (twice), $\alpha \pm \sqrt{2}\beta$ for the planar form (one allyl unit and one isolated orbital).) Notice that the two middle orbitals of the planar structure have the same energy. Thus with a total of four π electrons an orbital degeneracy occurs for the *HOMO*, and in a sense the geometry could be regarded as being Jahn-Teller unstable. The preferred structures as a function of electron count are given in Table 6.9.

Here lies then the origin of van't Hoff's rule which in this case tells us that the twisted structure should be more stable. It is a part of a much more general orbital result that when the collection of orbitals are half full of electrons then the most stable structure is that for which the orbital interactions are as similar

Table 6.9.

number of electrons	structure
2	D_{2h} $(1b_{1u})^2$ *vs.* $(1e)^2$
4	D_{2d} $(1b_{1u})^2(b_{3u})^2$ *vs.* $(1e)^4$
6	D_{2h} $(1b_{1u})^2(b_{3u})^2(b_{3g})^2$ *vs.* $(1e)^4(2e)^2$
8	D_{2d} $(1b_{1u})^2(b_{3u})^2(b_{3g})^2(2b_{1u})^2$ *vs.* $(1e)^4(2e)^4$

as possible. So here the doubled ethylene arrangement is more stable than the one involving one allyl and an isolated orbital.

6.18. First order perturbation theory tells us that the change in energy of an orbital as a result of an electronegativity perturbation is $\Delta e = c_i^2 \delta\alpha$, where c_i is the coefficient of the perturbed orbital in the unperturbed system and $\delta\alpha$ is the change in its Hückel α value. **6.13** shows the orbital energies and wavefunctions for the π manifold of benzene. The are given algebraically in the answer to Question 6.2. Using these values and noting that one component of the e_{1g} orbital (e_{1g}^b) has a zero coefficient at the carbon atom where the substitution occurs, $\Delta e(a_{2u}) = (1/6)\delta\alpha$, $\Delta e(e_{1g}^a) = (1/3)\delta\alpha$ and $\Delta e(e_{1g}^b) = (0)\delta\alpha$. This is shown schematically for a substituent less electronegative than carbon in **6.40**. Thus we expect to see three lines in a plot of ionization energy *versus* the p orbital energy of the substituent, with slopes close to 0.16, 0.33 and 0. In fact fitting of the data of **6.5** gives 0.16, 0.36 and 0.0.

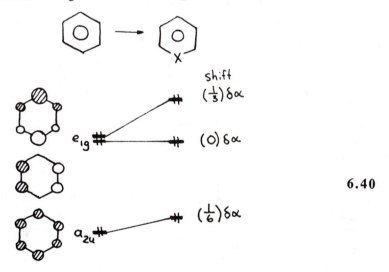

6.40

6.41

6.19. We found, using the ionization energies from Chapter 1 and a value of β of 5.0eV, an electronic energy of 107.06eV for the π energy of isomer **a** and 108.58eV for isomer **b**. Experimentally **b** is the only structure known; isomer **a** is not known. In fact setting the β values for B-B and N-N interactions at 4.5 and 5.5eV respectively does not change the energy of **a** at all, a result which is understandable from the theory of matrices. The qualitative explanation of the result is related to that considered in question 6.7. There the first order energy shift on substitution of the parent was sufficient to distinguish between the two isomers. The wavefunctions for the benzene molecule are given in the answer to Question 6.2 and **6.13**. However, evaluation of the first order shift, the sum $\Sigma_i c_{ij}^2 \delta\alpha$ for each orbital j (where the summation is over all of the atomic orbitals i), leads to identical (and zero) energy changes for the occupied orbitals in both cases. However the second order energy shift is given by an expression of the form $<i/\delta\mathcal{H}/j>/(e_j^{(0)}-e_i^{(0)})$, and for the mixing of the degenerate orbitals, $/i>$, $/j> = e_1, e_2$, this turns out to be larger for the alternating isomer **b** as shown in **6.41**, than for the isomer **a**. In this diagram we concentrate on the form of the numerator since it is the only part of this expression which will be different for the two isomers. We first evaluate the products of the orbital coefficients (neglecting normalization for simplicity) and then multiply these by the corresponding shifts $\delta\alpha$, positive for $C \rightarrow B$ but negative for $C \rightarrow N$. You can easily show, using the same approach, that the second order mixing for the non-degenerate orbitals is zero for isomer **a** but nonzero for isomer **b**. Thus for both sets of occupied orbitals pattern **b** is the favored one.

6.20. The odd electron in the radical anion must occupy one component of the e_{2u} orbital, but which one? It depends on the relative energy of the relevant π-type orbital on the substituent. For the electronegative CH_3 unit, a filled, deep-lying $CH_3(\pi)$ orbital interacts with one component of the e_{2u} orbital, $e_{2u}{}^a$, and

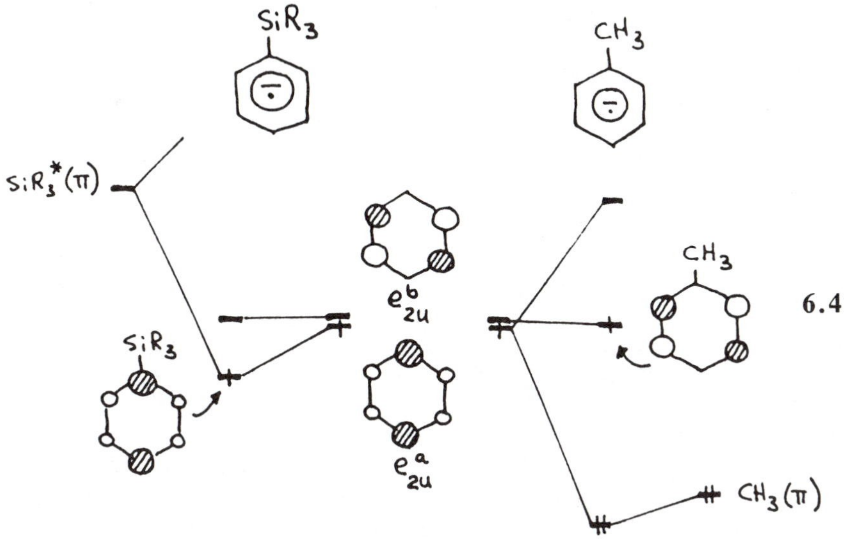

6.42

pushes it to higher energy leaving the e_{2u}^b orbital the lower energy of the two and the one which holds the odd electron as shown on the right side of **6.42**. The antibonding $CH_3^*(\pi)$ orbital lies too high in energy to be important. The converse is true for the SiR_3 substituent. Here the $SiR_3^*(\pi)$ is empty but importantly is energetically low-lying. It stabilizes the e_{2u}^a orbital on the benzene ring, and leads to single occupancy of this orbital in the radical anion as shown on the left side of **6.42**. The products of Birch reduction are easy to predict since hydrogen attacks those sites of highest electron density. The two products are shown in **6.43**.

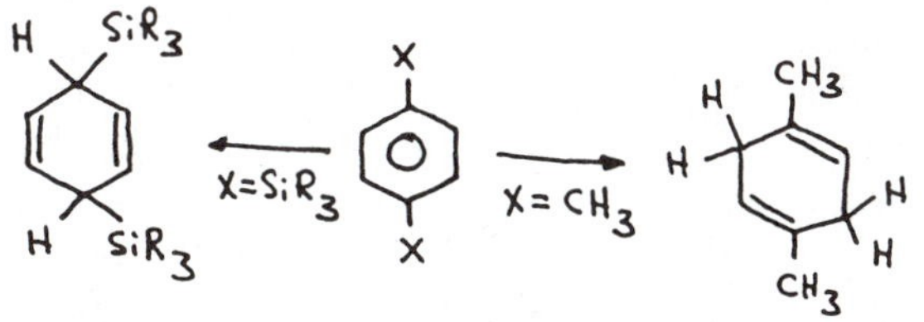

6.43

6.21. The longer central C-C distance is easy to understand. The *HOMO* of ethylene is a π-bonding orbital, and the loss of two electrons leads to removal of the C-C π bond. Accompanying its loss is the necessity for a planar N_2CCN_2 skeleton. The two ends of the molecule rotate around the C-C single bond to minimize steric interactions. Each CCN_2 unit is planar, as expected for a six electron AB_3 unit (e.g., BF_3) but the planar $CN(CH_3)_2$ units imply a significant π bonding network extending over the N-C-N unit. With three π orbitals to be considered and four electrons (the electrons which were originally lone pairs in the neutral parent) the π orbital picture is simply that of **6.44**. This shows good *CN* bonding (the system is isoelectronic with allyl anion); hence the shorter *CN* bond, with the π electron density residing primarily on the terminal nitrogen atoms. In the neutral molecule each nitrogen held a lone pair, thus in the cation electron density is primarily removed from the central

N—C—N

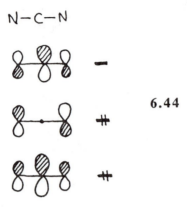

6.44

carbon atoms. Thus the major effect of removing electrons fron the *C-C* π bond is to free up a carbon $p\pi$ orbital which becomes involved in *N-C-N* π bonding.

6.22. There are several ways to construct the π orbitals for bismethylene-cyclobutadiene; two routes are shown here. In **6.45** notice that ψ_3 and ψ_4 are predicted to be degenerate. This 'accidental' degeneracy comes from the Hückel approximation, where only nearest neighbor overlap is taken into account. There is no reason from the symmetry of the carbon skeleton for it to come out this way. By including non-nearest neighbor interactions (those across the four-membered ring will be the largest on distance grounds) ψ_3 will lie slightly lower than ψ_4. The energy difference though will still be small. An alternative route is shown in **6.46**. Notice that in either case the energies of ψ_3 and ψ_4 clearly come out to be very close. Therefore the orbital occupation will be $(1b_{1u})^2(b_{3g})^2(2b_{1u})^2$ or $(1b_{1u})^2(b_{3g})^2(b_{2g})^2$ both of which lead to a ^1A_g state. Alternatively one could have the electronic configuration $(1b_{1u})^2(b_{3g})^2(2b_{1u})^1(b_{2g})^1$, which leads (*via* the direct product $b_{1u} \times b_{2g} = b_{3u}$) to $^1B_{3u}$ and $^3B_{3u}$ states.

6.23. This problem highlights the utilty of Wade's rules in predicting the geometries of cage and cluster molecules. Each *CH* unit uses a pair of electrons to form the extra-deltahedral *C-H* bond, leaving three electrons for skeletal bonding. Thus $C_5H_5^-$ has $(3\times5) +1 = 16$ and $C_5H_5^+$ $(3\times5) -1 = 14$ electrons involved in skeletal bonding. Since an *n*-vertex deltahedron is stable for $(n+1)$ pairs of electrons, $C_5H_5^-$ has an *arachno*-pentagonal bipyramidal geometry and $C_5H_5^+$ a *nido*-octahedral structure as shown in **6.47**.

6.24. Normally groups larger than hydrogen prefer to occupy the equatorial sites of cyclohexanes since here steric problems are reduced. However placing the hydroxy group in an axial position allows one of the lone pairs of the ring oxygen to lie in the same plane as the antibonding *C-OH* σ orbital. If the interaction between the two (shown in **6.48**) is sufficiently large then the axial

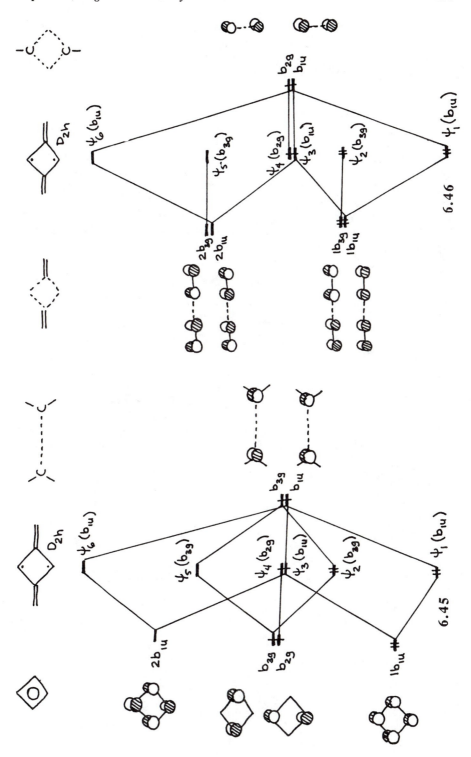

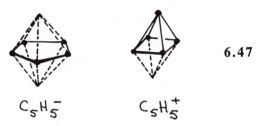

$C_5H_5^-$ $C_5H_5^+$ **6.47**

conformation will be preferred. This will be the case if the carbon substituent
is an electronegative one, such as *OH* (the case here) or *Cl* for example. This is
one of many types of structure where stereochemical preferences are set by
interactions which stabilize the *HOMO* after the largest bonding interactions
holding the atomic sleleton together have been accounted for. Electronically it
is a mechanism similar to that used to stabilize planar three (Question 4.11) and
four (Question 6.6) coordinate molecules, and has strong similarities to the
hyperconjugative mechanism of Question 6.4.

6.48

Chapter VII

Reactions

7.1. The replacement of CO by triphenylphosphine in $Ni(CO)_4$ follows a simple first order rate law, independent of the concentration of phosphine, whereas that for the isoelectronic compound $Co(CO)_3NO$ is much more complex. Why?

*7.2. A number of sulfides undergo nucleophilic substitution reactions of the type:

$$Nuc^- + R - S - X \rightarrow R - S - Nuc + X^-$$

To determine the most favored approach of the nucleophile, Nuc^-, towards the sulfide, choose the easiest situation where $Nuc^- = Cl^-$, $R = CH_3$ and $X = Cl$. Consider the C-S-Cl angle to be 90° in CH_3SCl. A polar coordinate system for the attack of Cl^- on CH_3SCl is shown in **7.1**. Note that ϕ can vary, in

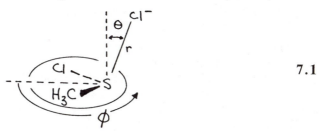

7.1

principle, from 0° to 360° where the dotted line at $\phi = 0°$ bisects the C-S-Cl angle. Choose a value of r so that there is a moderate amount of interaction between CH_3SCl and Cl^-.

(a) Use perturbation theory considerations to order the relative energies for the five positions of attack listed below. (Hint: be aware of the difference between CH_3 and Cl).

    1) $\phi = 135°$, $\theta = 90°$
    2) $\phi = 180°$, $\theta = 90°$
    3) $\phi = 225°$, $\theta = 90°$
    4) $\phi = 180°$, $\theta = 0°$
    5) $\phi = 180°$, $\theta = 45°$.

(b) Draw a hypothetical potential energy surface using a circular grid like that shown in **7.2**. Draw energy contours from ~1-10 (in hypothetical energy

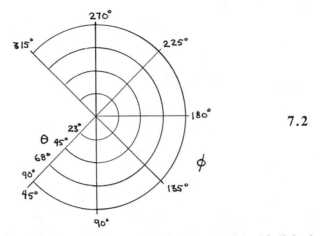

units) where 1 represents a lower potential energy than 10 (0 is the minimum) on the circular grid. Collect as much experimental data from the literature as possible to support your energy surface.

*7.3. The hydrogenation of olefins catalyzed by transition metal complexes is a very important industrial reaction. The initial step in the catalytic cycle is addition of H_2 to a square planar complex to form an octahedral metal

dihydride. There are two possible ways that H_2 can approach a d^8 ML_4 complex; these are shown in **7.3** and **7.4**.

(a) Show the most critical interactions between H_2 and the ML_4 complex for reactions **7.3** and **7.4** at the early stages of reaction, *i.e.*, when the orbital interactions between the reactants are strong but not much geometrical change occurs in the two molecules.

(b) Experimental evidence suggests that **7.3** is preferred over **7.4**. Determine the crucial orbital difference between the two geometries which will allow this to be so.

(c) For $ClIr(CO)(PPh_3)_2 + H_2$ there are two possible reaction paths that could be taken. These are shown in **7.5** and **7.6**. Notice that two different isomers are formed depending on the route taken. Experimentally, only one product is formed, that expected from **7.6**. On the basis of your answers to parts (a) and (b), show why the reaction path **7.6** is favored over **7.5** for $ClIr(CO)(PPh_3)_2$.

*7.4. The decomposition of PX_5 to $PX_3 + X_2$ is a reaction that has been known for sometime, particularly for $X = Cl$. The first reported example was by Mitscherlich, (*Ann. Phys. Chem.*, **29**, 221 (1883)). Several other papers on the reaction were published in the 19th century and many in the 1930's-1940's. However, the mechanism of this reaction is still not known with any certainty. The purpose of this exercise is to illustrate some of the problems associated with proposing a reasonable mechanism. First of all, the decomposition proceeds smoothly in nonpolar solvents which suggests that ionic intermediates $(PX_4^+ + X^-)$ are probably not formed. Secondly, no evidence has been encountered for the existence of free radicals. This implies that a concerted reaction, where P-X bond breaking occurs at the same time as X-X bond making, may be a possibility. Shown in **7.7** and **7.8** are two least-motion

7.7

7.8

ways that the decomposition can occur. Draw an orbital correlation diagram for each reaction path for $X = H$ and point out why both reaction mechanisms should require very high activation energies.

7.5. Explain why the rate of a typical Diels-Alder reaction increases as electron-withdrawing substituents are attached to the dienophile.

7.6. Construct an orbital correlation diagram for the reaction $He^+ + H_2$ to show how the 'expected' products (HeH^+ and H) are not observed, but the 'unexpected' ones (H^- and HeH^{2+}) are. Remember He is more electronegative than H.

7.7. Molecules such as nitrogen and acetylene are able to quench excited states of the alkali metals. These excited states contain an electron in the valence p orbital rather than the valence s orbital appropriate for the ground state. If $M*$ represents such an excited alkali metal atom, then the reaction is simply $M* + X_2 \rightarrow M + X_2$ and the electronic excitation energy appears in other degrees of freedom of the X_2 molecule. The way this occurs may be modeled using the simple case of $X_2 = H_2$. Construct a molecular orbital diagram for M coordinated to H_2 in a sideways or η^2 mode and show, by considering their bonding characteristics, how the various molecular orbitals change in energy as the M - H_2 distance (r) changes. Hence show the existence of an exciplex (a complex stable in an excited state but not in the ground state) and devise a model for the quenching.

7.8. Inner-sphere electron transfer reactions between transition metal ions in solution are though to proceed *via* the path shown in **7.9**. Construct a molecular orbital diagram for a symmetrical L_5M-X-ML_5 complex and use

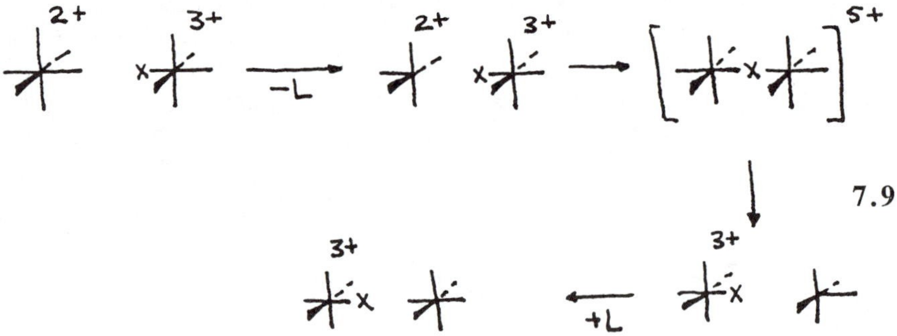

this to build a correlation diagram for the atom transfer process shown in **7.9**, namely $L_5M...X\text{-}ML_5 \rightarrow L_5M\text{-}X\text{-}ML_5 \rightarrow L_5M\text{-}X...ML_5$. Hence show how the redox process occurs for the case of (high-spin) Cr^{III} and Cr^{II}.

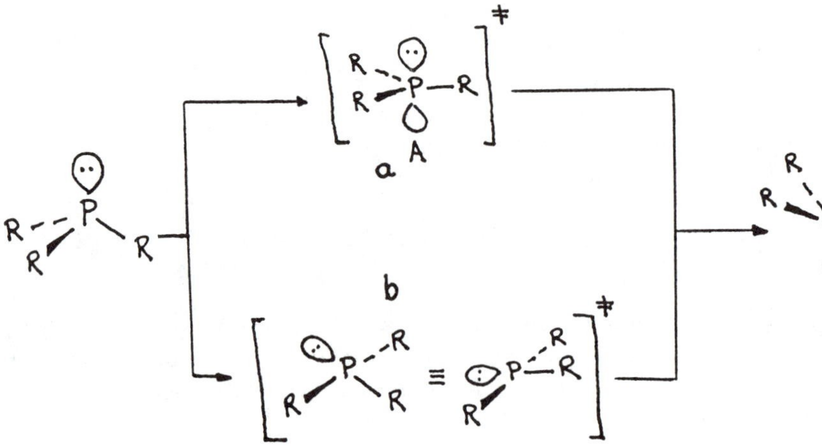

7.9. A new mechanism has been proposed for pyramidal inversion in phosphines. Shown in **7.10** is the traditional mechanism which proceeds via a D_{3h} transition state labeled **a** and the new mechanism which proceeds via a C_{2v} transition state labeled **b**. There is some experimental evidence that suggests that **b** can become energetically more viable than **a**. However, the most direct evidence comes from a series of molecular orbital calculations. Listed in Table 7.1 are the calculated activation energies for each process in a series of phosphines. Show why the Transition State **a** is favored over **b** for PH_3 although **b** is the favored transition state in PF_3.

Table 7.1

PR_3	ΔE_a	ΔE_b (kcal/mole)
PH_3	38	159
PH_2F	60	127
PF_2H	102	66
PF_3	125	69

7.10. Develop an orbital correlation diagram for the attack of a carbene on an olefin, and show that this reaction as drawn in **7.11** is symmetry forbidden. In

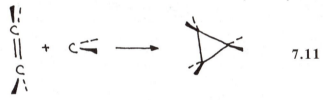

7.11

fact it is a fairly common reaction. How should you modify your scheme to allow for this?

7.11. **7.12** shows the results from a series of structural studies of solids containing molecules where there is a relatively close approach between a

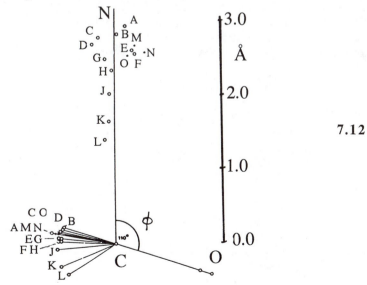

7.12

carbonyl and a nucleophilic group. Notice that the observed points fall in a well-defined pattern. These results have been interpreted as mapping out the pathway of nucleophilic approach to a carbonyl. Examine the orbital interactions between the two molecules as a function of the angle ϕ, and provide a rationalization of the observed crystallographic data.

7.12. Use symmetry conservation rules to decide whether the addition of Cl_2 to ethylene is thermally allowed or not.

Answers

7.1. $Ni(CO)_4$ and $Co(CO)_3NO$ are both eighteen electron species, a number reached by regarding CO as a two-electron donor and the linear NO molecule as a three-electron donor. Coordination of an extra ligand such as a phosphine to $Ni(CO)_4$ would lead to an energetically unfavorable twenty-electron species. Thus substitution processes for this species should involve ligand loss followed by coordination of phosphine. Thus a first order rate law is understandable. In the second case the molecule may take advantage of the well-known variability in the nitrosyl geometry. If the MNO angle is 180° then the ligand is a three-electron donor, but if the angle is much lower, such that a lone pair develops at nitrogen, then the ligand is regarded as a one-electron donor. Thus there is probably an energetically low-lying geometry for this molecule, which is still an eighteen electron one, of stoichiometry $Co(CO)_3NOPR_3$ where the MNO unit is non-linear. This is a very general result found in the reactions of many organometallic complexes containing ligands with variable coordination geometries. For example a coordinated η^6-benzene molecule may slip to an η^4 geometry and thus open up a vacant site at the metal as the nucleophile approaches in an exactly analogous way to that of NO.

7.2. The key to this problem is to maximize *HOMO-LUMO* interactions while at the same time minimizing two-orbital, four-electron repulsions. Since the orbitals on Cl^- are completely filled we only need to worry about maximizing interactions with the *LUMO* on CH_3SCl and minimizing the interactions

$-\ 2b_2$

$-\ 3a_1$ **7.13**

$+\!\!\!+\ b_1$

$+\!\!\!+\ 2a_1$

between the p $AO's$ on Cl with filled orbitals on CH_3SCl. The important orbitals of CH_3SCH_3 are shown in **7.13**. Certainly Cl is much more electronegative than $-CH_3$. The b_1 and $2a_1$ orbitals will not be affected much by this electronegativity perturbation, but there will be substantial intermixing of $2b_2$ and $3a_1$. The rules of perturbation theory allow us to decide whether these energy levels go up or down in energy as a result of this substitution. Writing $C_\alpha^0(abc)$ as the coefficient of the atomic orbital (α) whose H_{ii} value will be changed (by $\delta\alpha$) in the unperturbed (^{0}) molecule in the molecular orbital (abc) and recalling that $\delta\alpha = (-)$:

For $3a_1$: $\quad e^{(1)}(3a_1) = (C_\alpha^0(3a_1))^2\delta\alpha = (+)^2(-) = (-)$

$$e^{(2)}(3a_1) = (C_\alpha^0(3a_1)C_\alpha^0(2b_2)\delta\alpha)^2/(e^0(3a_1)-e^0(2b_2)) = (+)/(-) = (-)$$

$$t^{(1)}(2b_2,3a_1) = C_\alpha^0(3a_1)C_\alpha^0(2b_2)\delta\alpha/(e^0(3a_1)-e^0(2b_2)) = (+)(+)(-)/(-) = (+)$$

For $2b_2$: $\quad e^{(1)}(2b_2) = (C_\alpha^0(2b_2))^2\delta\alpha = (+)^2(-) = (-)$

$$e^{(2)}(2b_2) = (C_\alpha^0(3a_1)C_\alpha^0(2b_2)\delta\alpha)^2/(e^0(2b_2)-e^0(3a_1)) = (+)/(+) = (+)$$

$$t^{(1)}(3a_1,2b_2) = C_\alpha^0(3a_1)C_\alpha^0(2b_2)\delta\alpha/(e^0(2b_2)-e^0(3a_1)) = (+)(+)(-)/(+) = (-)$$

The perturbed orbitals then look like **7.14**.

The p A.O. on Cl^- which is oriented along r should maximize its overlap with σ^*_{S-Cl} and minimize its overlap with n_π and n_σ.

At (1) $\phi = 135°$, $\theta = 90°$. Maximum overlap with σ^*_{S-Cl}, small overlap

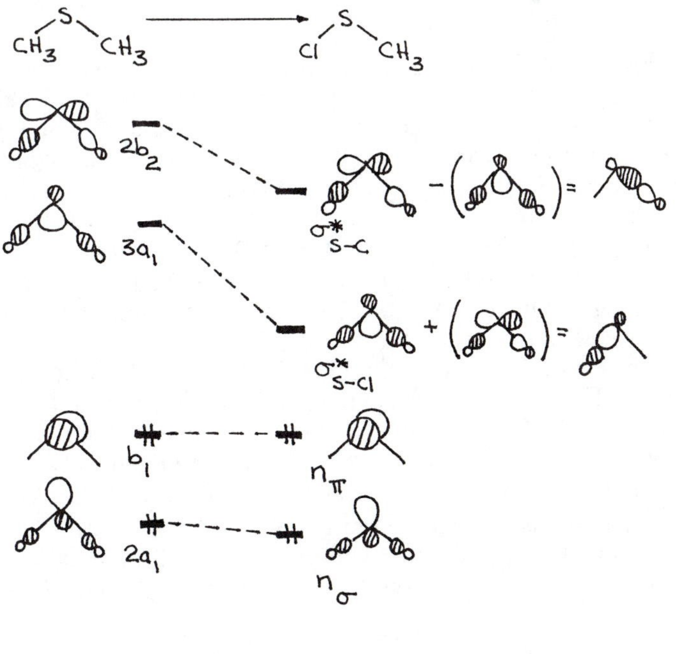

7.14

with n_σ, small overlap with $\sigma*_{S-C}$ and 0 overlap with n_π. This is the best orientation.

At (2) $\phi = 180°$, $\theta = 90°$. Small overlap with $\sigma*_{S-C}$ and $\sigma*_{S-Cl}$, maximum overlap with n_σ, 0 overlap with n_π.

At (3) $\phi = 225°$, $\theta = 90°$. Maximum overlap with $\sigma*_{S-C}$, small overlap with $\sigma*_{S-Cl}$ and n_σ, 0 overlap with n_π. Since $\sigma*_{S-C}$ lies higher in energy than $\sigma*_{S-Cl}$ this is not as good as (1).

At (4) $\phi = 180°$, $\theta = 0°$. Maximum. overlap with n_π, ~0 overlap with the rest.

At (5) $\phi = 180°$, $\theta = 45°$. Strong overlap with *both* n_π and n_σ, ~0 overlap with the rest.

Thus the relative ordering of stability for a given r is: 1>3>>2>4>5.

(b) A hypothetical potential energy surface might look like that in **7.15**.

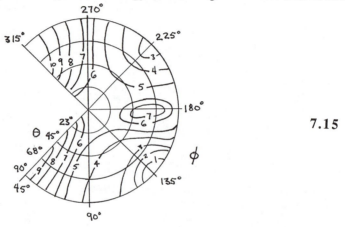

7.15

There are a number of pieces of experimental evidence for this type of potential energy surface. The most direct comes from x-ray crystallography. There are many structures containing two-coordinate sulfur of type Y-S-Z, where Y and Z can be a variety of groups. These structures frequently show close intermolecular contacts of the *Nuc: ...S* sort. The geometrical location of the nucleophiles relative to the Y-S-Z framework are plotted in **7.16**. (The plot is from *J. Amer. Chem. Soc.*, **99**, 4860 (1977).) The open circles are the important ones (closed circles represent electrophiles attacking Y-S-Z). Notice that they cluster around $\theta = 90°$, $\phi = 130°$ just as we predicted. Notice that there are very few points in the $\theta = 0 - 90°$, $\phi = 180°$ region which we predict to be of high energy. Notice in the plot above ϕ runs from 0° to 180° (not 360° in ours). This is because the authors did not discriminate between Y and Z in the set of Y-S-Z compounds. Finally, note that for the case of an electrophile attacking CH_3SCl, we would want to maximize the interaction between the *LUMO* of the electrophile and n_σ/n_π on CH_3SCl. This would lead to favored orientations of attack corresponding to $\phi = 180°$, $\theta = 90°$-0°. The solid circles show this to be the case with $\phi = $ ~160°-180°, $\theta = 0°$-30°.

Another piece of information is given by the fact that the transition state

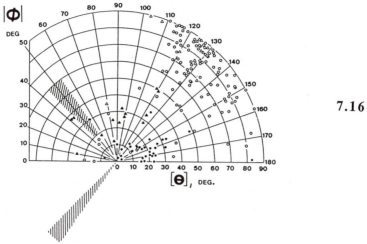

7.16

for nucleophilic attack is the 10 electron $CH_3SCl_2^-$ species. *VSEPR* tells us that this is based on a trigonal bipyramid, of which, there are five possible isomers (**7.17**). We know from site preferences of a trigonal bipyramid that

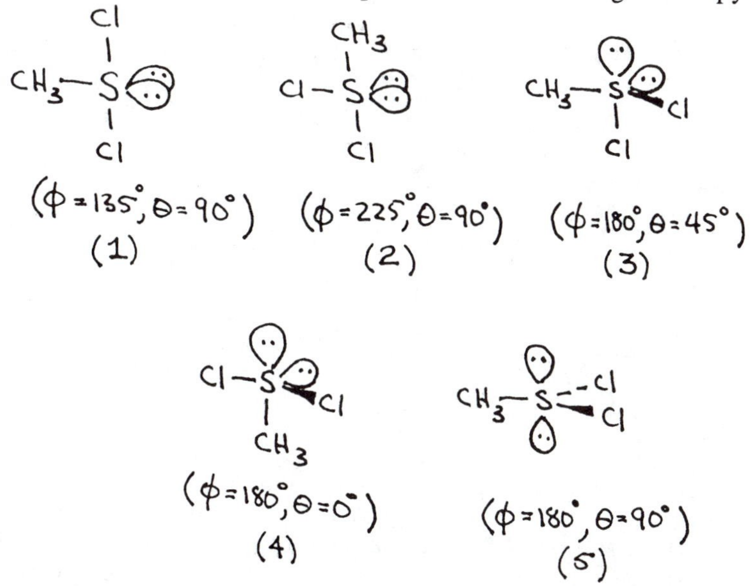

$$CH_3-S\overset{Cl}{\underset{Cl}{|}}\ominus$$

$$(\phi=135°, \theta=90°)$$
$$(1)$$

$$Cl-S\overset{CH_3}{\underset{Cl}{|}}\ominus$$

$$(\phi=225°, \theta=90°)$$
$$(2)$$

$$CH_3-\overset{\cdot\cdot}{S}\underset{Cl}{|}Cl$$

$$(\phi=180°, \theta=45°)$$
$$(3)$$

7.17

$$Cl-\overset{\cdot\cdot}{S}\underset{CH_3}{|}Cl$$

$$(\phi=180°, \theta=0°)$$
$$(4)$$

$$CH_3-\overset{\cdot\cdot}{\underset{\cdot\cdot}{S}}\overset{-Cl}{\underset{Cl}{}}$$

$$(\phi=180°, \theta=90°)$$
$$(5)$$

electronegative groups prefer the axial positions, electropositive ones, the equatorial positions. In this case the electronegativity of the groups are in the

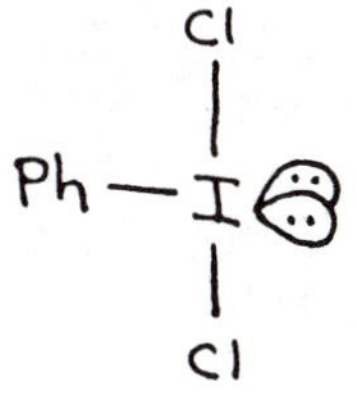

$$Ph-I\overset{Cl}{\underset{Cl}{|}}\ominus$$

7.18

order $Cl > CH_3 >>$ *lone pairs.* Thus the stability should be (1) > (2) > (3) > (4) > (5). Experimentally, the compound that comes closest to our situation is *Ph-ICl₂*, with the structure shown in **7.18** corresponding to (1).

7.3   (a) The interaction diagram between H_2 and the square plane in the geometry of **7.3** is shown in **7.19**.

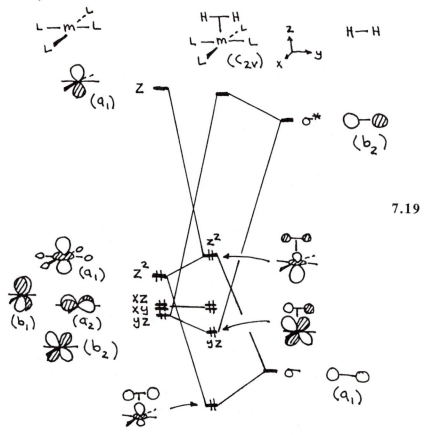

7.19

The interaction diagram for the geometry of **7.4** is shown in **7.20**.
(b) There are two major differences between the pathways. In **7.3** there is interaction between yz and σ^* which stabilizes the occupied yz orbital. In **7.4** there is no analogous stabilization of a metal-centered d orbital. In **7.3**, although there is some repulsion between $H_2 \sigma$ and z^2 (both orbitals are occupied) this is mediated to a large extent by interaction with the metal z orbital. In **7.4** there is a similar situation, but here the $H_2 \sigma^*$ orbital can also mix into z^2 to relieve this repulsion. For **7.3** to be more favorable the stabilization of yz by $H_2 \sigma^*$ must be greater than the stabilization of z^2 by H_2 σ^* in **7.4**. Similar arguments to those we have just used should also apply to the case of I_2 attack. In this case there is a molecule, $[Pt^{II}I\{C_6H_3(CH_2NMe_2)_2\text{-}o,o'\}(\eta^1\text{-}I_2)]$, which has the geometry shown in **7.4** but with I_2 in the place of H_2.

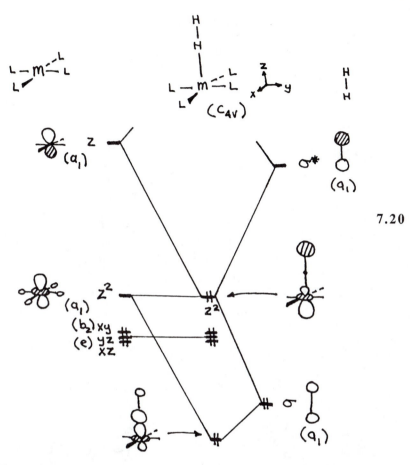

7.20

(c) If one ignores π effects associated with the *Cl* and *CO* ligands then reactions **7.5** and **7.6** should occur with approximately the same activation energy. However there is an antibonding interaction (**7.21**) between the H_2 σ* orbital and the *Cl z* contribution to 'yz' which leads to a smaller overlap, and thus smaller stabilization, than that between *xz* and H_2 σ*. Recall that the carbon *z* character in *xz* is very small.

vs

III   III   **7.21**

vs

7.4. A correlation diagram for the process in **7.7** is shown in **7.22**. It may be readily constructed from the orbital pictures for the three different molecules and taking into account the symmetry of the system. The problem with this

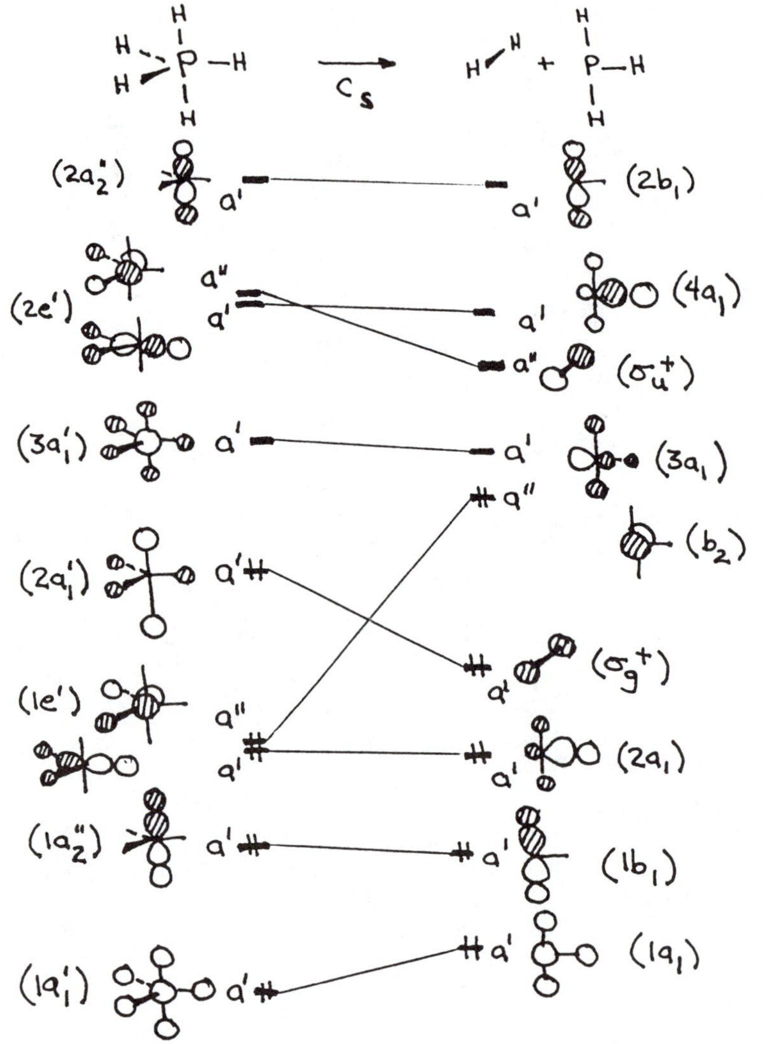

7.22

reaction path is that one component of the *1e'* set must rise to a very high energy and become the b_2 orbital in a C_{2v} PH_3 molecule. (See Question 4.37.) C_{2v} PH_3 lies 159 kcal/mol *above* the energy of C_{3v} PH_3. The only way to get

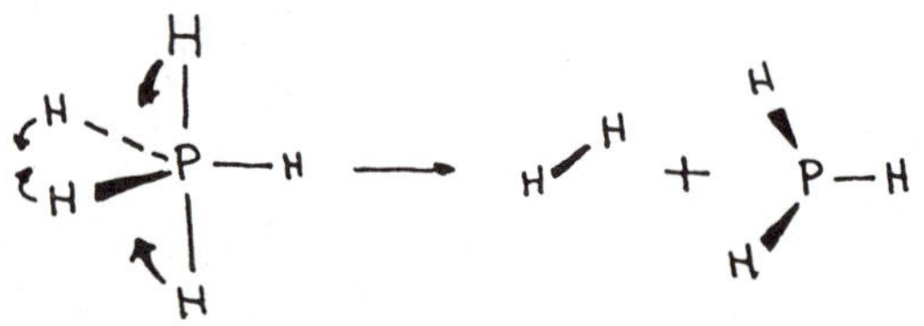

7.23

around this is to use the more complex motion of **7.23**. This particular component of the *1e'* set then ultimately becomes the lone pair hybrid orbital of *PH₃*.

The orbital correlation diagram for **7.8** is shown in **7.24**. The problem with this pathway is that the *2a₁'* orbital of *PH₅* should correlate with the σ_u^+ orbital of H_2 and *3a₁'* with σ_g^+. There is a weakly avoided crossing between

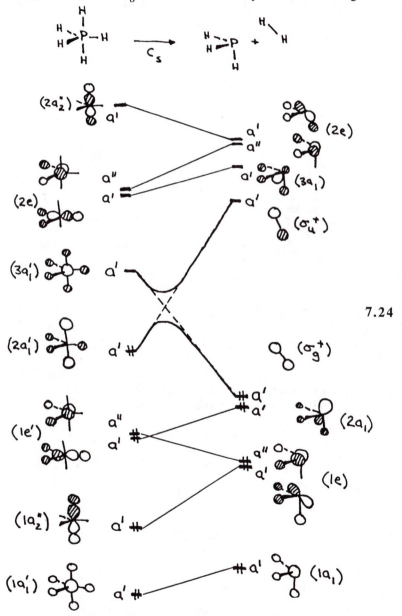

7.24

the two *a'* levels but the energy of *2a₁'* must rise substantially to give a very high activation barrier.

7.5. The important orbital interaction between diene and dienophile in the case where the dienophile has a low-lying *LUMO* is that shown in **7.25** and is the interaction between diene *HOMO* and dienophile *LUMO*. Attaching electron-withdrawing substituents to the dienophile results in a lowering of the dienophiile energy levels. This *HOMO-LUMO* gap becomes smaller and the activation energy decreases. Experimentally this shows up quite dramatically in reaction rates. With butadiene, addition of ethylene gives a 70% yield in 17 hrs at 165°C and 900 atmospheres pressure, but with maleic anhydride a 100% yield is found after 24 hrs and 20°C.

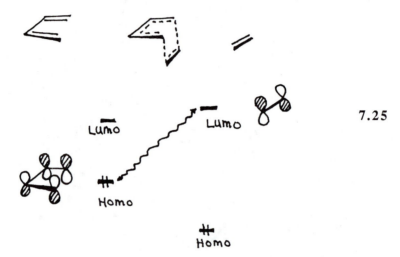

7.25

7.6. **7.26** shows the orbital correlation diagram for the process. Since *He* is much more electronegative than *H*, its *1s* level lies deeper than the H_2 bonding orbital at the left hand side of the diagram. In the center are the orbitals of the HeH_2^+ unit. In the deepest-lying orbital the helium *1s* orbital contribution is largest on electronegativity grounds. This orbital has to correlate with the deepest lying orbital of the product, namely the *H-He* bonding orbital. The middle orbital of the center panel must correlate with the energetically closest

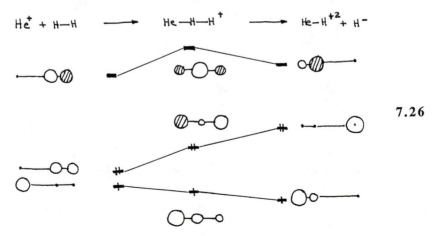

7.26

orbital at the right hand side of the diagram, namely a hydrogen $1s$ orbital. The overall result, since the deepest lying orbital is only singly occupied, is to produce an *HeH* species with only one electron, namely HHe^{2+}.

7.7. The molecular orbital diagram of **7.27** is a simple one to derive. The σ bonding orbital of H_2 may interact only with the s and p_z orbitals of the alkali metal by symmetry. The lowest member of the trio is M-H_2 bonding, the other two antibonding. Because of the large s/p energy separation the metal p orbital is probably only pushed up a little in energy. The lowest member of the p orbital set is stabilized by interaction with the σ^* orbital on H_2. One p orbital component remains unchanged in energy. In the electronic ground state, with three electrons, two remain in the σ orbital of H_2, bonding between the two partners, but one is located in the metal s orbital, antibonding between them. It is not usually possible to decide in general whether such three electron-two orbital problems are overall bonding or antibonding. In this case we assume that overall it is weakly bonding. The energy of this state (I) as a function of distance therefore shows a shallow minimum (**7.28**). In the first excited state (II) the electron in this antibonding orbital is removed and placed in an orbital which is bonding between the two partners. A minimum in the $E(r)$ plot is thus to be expected. This is the exciplex. The two curves probably cross with decreasing r, thus allowing a smooth transition from the excited state to the ground state. The excess electronic energy is thus converted into translational energy of the two partners. (In practice the quenching of excited alkali metal atoms by acetylene is more complex. For example if $D^{13}C\equiv^{12}CH$ is used then $D^{12}C\equiv^{13}CH$ is found at the end of the process. One of the problems of quenching by hydrogen is that of fission of the H-H bond. For more

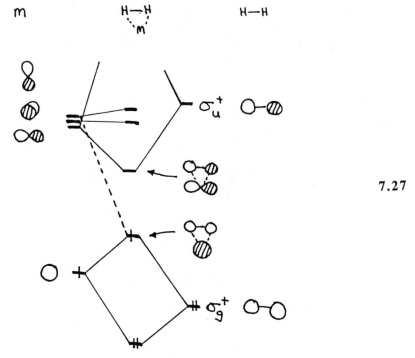

7.27

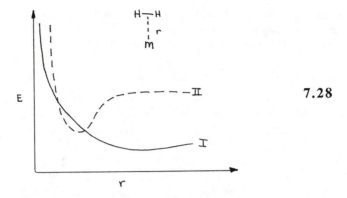

7.28

information see: *Chem. Phys.,* **143**, 39, (1990).)

7.8.   The molecular orbital diagram for the symmetrical L_5M-X-ML_5 complex is readily derived from the orbitals of two ML_5 units plus those of a bridging X atom as in **7.29**. Notice that the in- and out-of-phase combinations of the z^2 orbitals are not degenerate, being split apart in energy by different interactions with the X s and p orbitals. For the asymmetric $L_5M...X$-ML_5 complex the orbitals of the left hand system resemble those of the five-coordinate ML_5 molecule, whereas those of the right hand atom resemble those

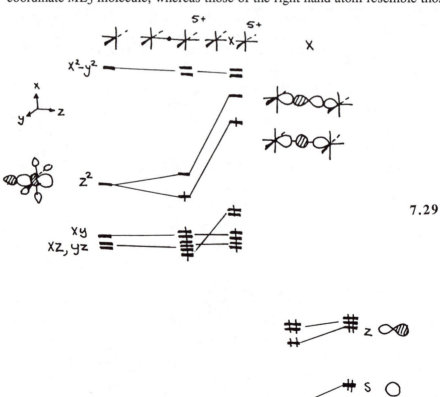

7.29

7.30

of an octahedral ML_6 complex and thus the correlation diagram is easy to derive (**7.30**). The single electron in the z^2 orbital is thus smoothly transferred from one metal atom to another as the atomic orbital composition of the *HOMO* gradually changes on moving the atom from one side of the bridge to the other. For more details see: *Inorg. Chem.* **17**, 2537 (1978).

7.9.   The important point here is to evaluate the energies of transition state **a** relative to **b** as a function of the electronegativity of *R*. From Question 4.20 we can understand how the inversion barrier *via* the D_{3h} structure increases with the electronegativity of the ligands. As the electronegativity increases $2a_1$ drops to lower energy and thus the second order Jahn-Teller destabilization of the the D_{3h} structure increases. It is the relative stability of the C_{2v} geometry which bears investigation. **7.31** shows a Walsh diagram for the distortion from **a** to a T-shaped structure, **b**. The D_{3h} geometry is normally the more stable structure since the loss of bonding in $2a_1$ outweighs the gain in bonding in $1b_1$. (The most stable electronic arrangement is always that where the interactions are as equal as possible. See Questions 4.14, 4.32.) Therefore, a higher activation energy will be required to go through **b** compared with **a**. This is consistent with the results for PH_3. In PF_3 all MO's except a_2'' will go down in energy. This is shown in the Walsh diagram by dashed arrows (Question 4.7(c)). This effect can be large enough so that, as shown in **7.31** the $3a_1'$ orbital can lie at a lower energy than b_2. If this occurs, then **b** can be at lower energy than **a**, and this apparently is the case for PHF_2 and PF_3.

7.10. The molecular orbital diagram appropriate for attack of a carbene on an olefin in the geometry shown is derived in **7.32**. The result is initially a strong two-orbital four-electron repulsion between the occupied olefin π orbital and

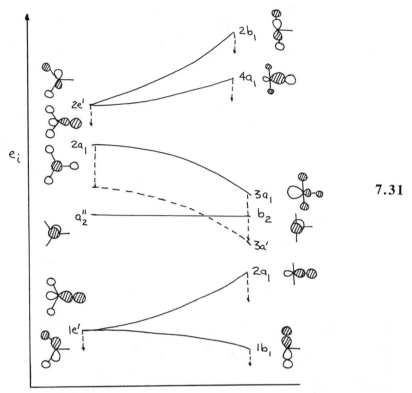

7.31

the carbene lone pair, leading to an energetically unfavorable approach. At closer distances, and therefore stronger interactions this geometry is fine since the π bonding and σ antibonding orbitals have now crossed. There is an interaction between the carbene π orbital and the olefin π^* orbital, but this is energetically unimportant since no electrons occupy these orbitals. However rotation of the carbene *in the initial stages of reaction* leads to an electronically much more favorable state of affairs as shown in **7.33**. Here there are two

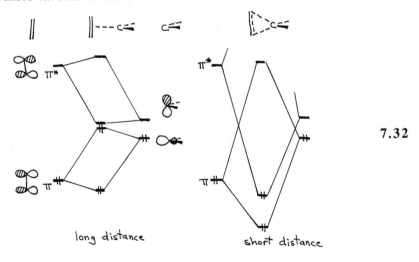

long distance

short distance

7.32

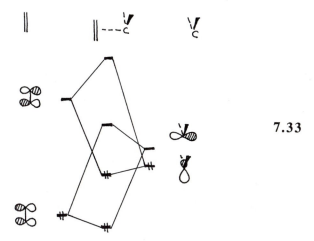

7.33

two-electron stabilizing interactions. Thus, we would predict that the energetically favorable route for this attack would be a corkscrew one, associated with rotation of the carbene through 90° on its approach.

7.11. **7.34** shows the interaction of the *HOMO* of an entering nucleophile wth the *HOMO* and *LUMO* of the carbonyl group. Interaction **a** is a two center-four electron one and is destabilizing. Interaction **b** is a two center-two electron one and is stabilizing. Thus the actual approach geometry will be the one that balances the stabilizing and destabilizing effects of the two

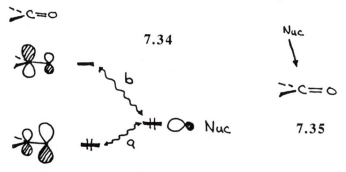

7.34

7.35

interactions. Attack will be at the carbonyl carbon atom, whose $p\pi$ orbital coefficient is largest in the *LUMO*. Attack on the carbon atom from both the 'on top' geometry and from the oxygen side is disfavored because of the destabilizing effect of interaction **a**. The most favored approach will then be that of **7.35**, a result in accord with the picture of **7.12**.

7.12. **7.36** shows an orbital interaction diagram for the sideways approach of two ethylene units. For the electron configuration appropriate for two ethylene molecules ($\psi_1^2\psi_2^2$), the result· is a two center-four electron one and is energetically destabilizing. For the electron configuration appropriate for a pair of ethylene and chlorine molecules, there is also a stabilizing interaction which arises from double occupation of ψ_3. The ethylene dimerization is thus

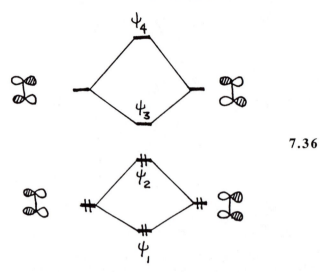

7.36

thermally forbidden, but the C_2H_4/Cl_2 reaction thermally allowed. Notice that the sideways approach of two ethylenes with two electrons less, *i.e.*, the $(\psi_1{}^2)$ configuration, would be expected to be stabilising. A series of hydrocarbons, called *pagodanes*, can be oxidised by two electrons to give such a species (*J. Amer. Chem. Soc.*, **110**, 7764 (1988)).

Chapter VIII.

Solids

*8.1. A solid, with an essentially infinite collection of orbitals, may be viewed as a very large cyclic molecule as in **8.1**. This means that we can use the group theoretical tricks learned for cyclic organic molecules to construct an 'energy band' of orbitals for the solid. (a) Use simple Hückel theory to construct such an energy band. What are the energies (in terms of the α and β parameters)

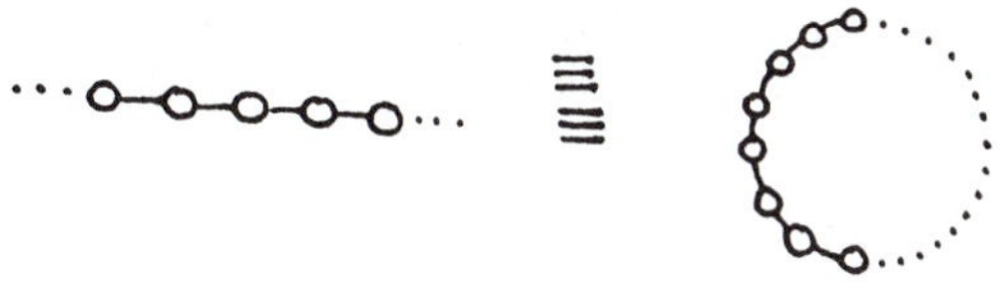

8.1 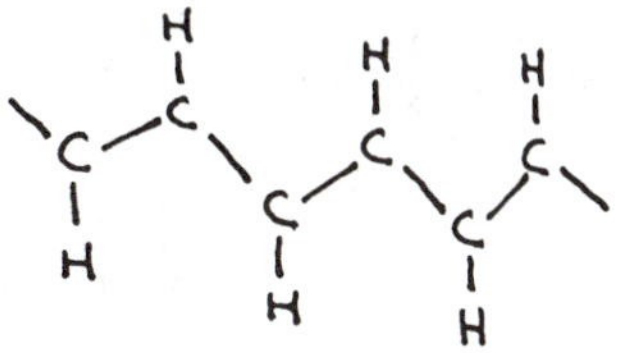   **8.2**

and orbital description at the very top and very bottom of the band? (b) What is the electronic configuration of the π manifold of polyacetylene, **8.2**. (c) What important physical property should this system have (it has actually been made and characterized) if it possesses this structure?

8.2. The species $K_2Pt(CN)_4$ has the structure shown in **8.3**. The energy band associated with the metal z^2 orbital will look very much like what you drew in

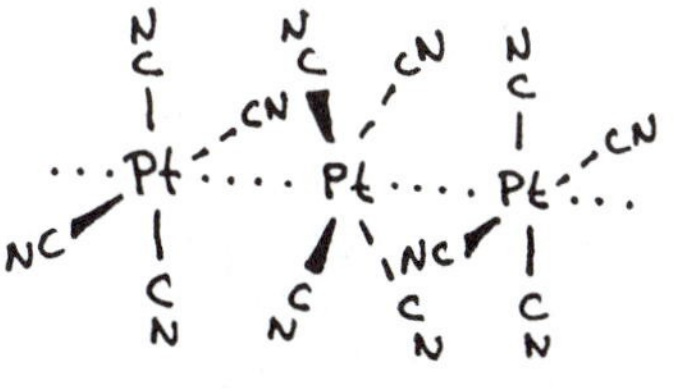

8.3

Question 8.1, except that instead of a $C\ 2p\pi$ orbital the $Pt\ z^2$ orbital is used as a basis. (a) Draw a rough molecular orbital diagram (for the d orbital region only) for the square-planar $Pt(CN)_4^{-2}$ unit. How many d electrons are located in the z^2 orbital? (b) Show how this orbital is broadened into a band in the solid. How many electrons per z^2 band per unit cell are there now? (c) The material can be doped with bromine to give a solid with a stoichiometry $K_2Pt(CN)_4Br_\delta$ where δ is about 0.3. How many electrons in the z^2 band now? Is the material expected to be an insulator or a metal? (d) By examining the nature of the orbitals (*i.e.*, their bonding or antibonding character) decide how the *Pt-Pt* distance will vary with δ.

*8.3. Explain why in **8.4**, system (a) is an antiferromagnetic insulator with equal metal-bridge distances but system (b) is an insulator with unequal metal-bridge distances. Here N represents a nitrogen-containing donor. The material is of stoichiometry $[MN_4X]^{2+}$ where X = Halogen.

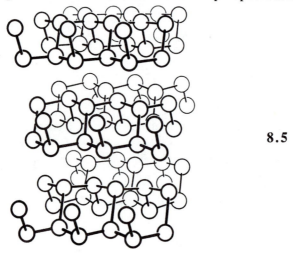

8.4

8.4. Assume that the band structure of the simple cubic structure can be represented by three $p\sigma$ bands appropriate for the three perpendicular (x, y, z) sets of p orbitals. Ignoring π-interactions and leaving the s orbital as a repository for two electrons generate the band structure for such a system. Hence explain the observed structure of black phosphorous shown in **8.5**.

8.5

*8.5. Take a one-dimensional chain of orbitals, for example the π orbitals of polyacetylene of Question 8.1. First generate the energy bands for a geometry where the *C-C* distances are equal, and then repeat the calculation for the case where the bonds alternate in length. Describe this electronic situation by using two different interaction integrals (β_1 and β_2) between adjacent orbitals to mimic this bond length difference. Would polyacetylene with this geometry be a metal or an insulator?

8.6. The structure of the mineral rutile, TiO_2, is shown in **8.6**. Draw out the expected form of the electronic density of states (*DOS*) and the Crystal Orbital Overlap Population (*COOP*) curve for the *Ti-O* overlap population. In your drawings, indicate the position of the Fermi level. Note: Make the energy scale for the *DOS* identical to that in the *COOP* curve and assume that the *Ti-O* distances are all equal and the *Ti-Ti* distances are long.

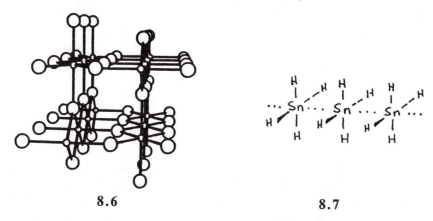

8.6   **8.7**

8.7. Consider a hypothetical chain of square planar SnH_4 units, as shown in **8.7**. Take the $Sn...Sn$ distance to be 2.90Å, slightly longer than a typical $Sn-Sn$ single bond. a) Draw out all eight molecular orbitals for the SnH_4 unit cell. Order them in energy, show the electron occupancy, and give the correct Mulliken symbol for each molecular orbital. b) Draw out the expected band structure for this material, showing the orbital phases and composition at the k = 0, π/a points for each band (use just three unit cells).

8.8. A number of solid-state compounds exist with the chemical formula $AB_{n-1}M_nO_{3n+1}$. Here A and B are very electropositive cations and M is a transition metal. They are technologically interesting because of their ferroelectrical properties and potential use as cation transport materials. The most simple member of this class has the stochiometry Sr_2TiO_4. The TiO_4 unit consists of a vertex shared sheet of octahedra. Above and below the sheets sit

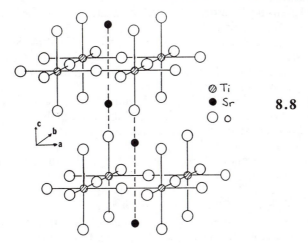

⌀ Ti
● Sr
○ O

8.8

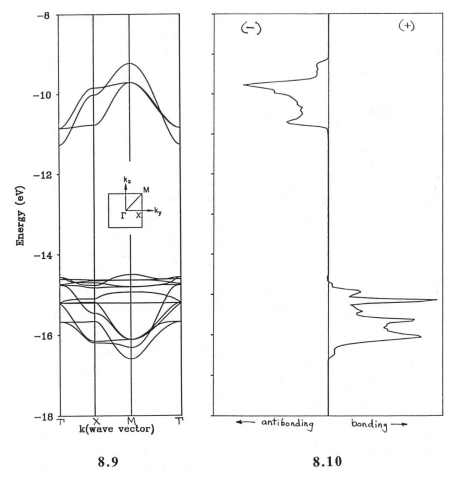

8.9   **8.10**

the *Sr* atoms. Thus, there are alternating sheets of TiO_4 octahedra and *Sr* atoms as shown in **8.8**.

(a) Roughly sketch the Electronic Density of States (*DOS*) for this material and describe their composition.

(b) **8.9** shows the band structure for the region from -18eV to -8eV and **8.10** shows the Crystal Orbital Overlap Population (*COOP*) curve for the *Ti-O* overlap population. Describe the composition of each group of bands and mark the expected position of the Fermi level.

(c) Draw the shape of the orbitals at Γ, *X* and *M* for the three bands in the energy range from -11.3eV to -9.0eV. Describe why each band goes up in energy or remains at constant energy as *k* changes.

*8.9. There are two forms of *PbO*. One of these, the so-called α form, is shown in **8.11** in two different perspectives:

(a) Notice that there are layers of *PbO*. Let us first just take one layer. The unit cell is Pb_2O_2 (doubled from *PbO* since *Pb* atoms are above and below the oxygen plane). The band structure for the region -17.5eV to -11.5eV is shown in **8.12**. All bands are fully occupied. Draw the basic shape for the bands

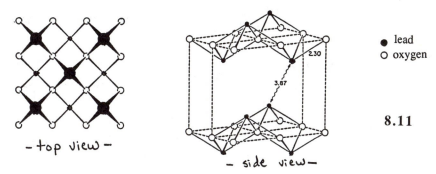

8.11

— top view —

— side view —

● lead
○ oxygen

labeled $3a_1$ and $4a_1$ at the Γ point. Hint: consider a square pyramidal (C_{4v}) AH_4 molecule where the hydrogens are much more electronegative than A. (b) The *COOP* curves for the intralayer *Pb-O* (solid line) and interlayer *Pb-Pb* linkages are shown in **8.13** for the full three dimensional (many layers) material. Explain the shape of both curves. (c) Notice in the energy interval from -13.5eV to -10.5eV in **8.13** that the amount of bonding *Pb-Pb* overlap population is greater than the antibonding (the total integral under the curve). This is not expected at a first glance. Why? Suggest a reason why this occurs.

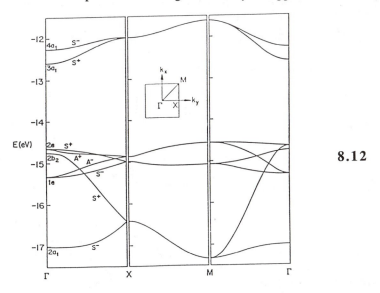

8.12

8.10. $CdIn_2Se_4$ is often described as having a defect sphalerite (ZnS) structure, namely one where some of the sites, occupied in ZnS are empty. This is shown in the pair of pictures of **8.14**. The formula is often written as $CdIn_2\square Se_4$ where $\square_x$ represents x vacant sites, to emphasise this relationship. Provide an orbital explanation for the presence of defects in this structure and make an analogy with the structures of *closo* and *nido* boranes, recognizing that a *nido* structure is a 'defect' *closo* one. Extend your reasoning to determine x in the structural formula of the solid described as $CuIn_2\square_xSe_3Br$.

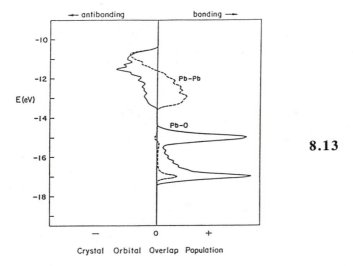

8.13

Crystal Orbital Overlap Population

8.11. The structure of the so-called '2-1-4' superconductor, $La_{2-x}Sr_xCuO_4$ is very similar to that shown in **8.8**, except that the axial Cu-O distances perpendicular to the sheet are quite long (~2.45Å). Rationalize the observation that as x increases the copper-oxygen distance in the CuO_2 plane decreases. (1.9035(1)Å for $x = 0$, 1.8896(1)Å for $x = 0.15$.)

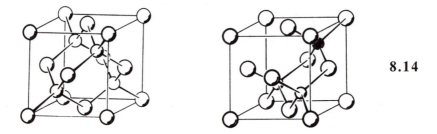

8.14

8.12. **8.15** shows a plan of the MB_4 structure. Convince yourself that this structure consists of sheets of B_6 octahedra linked by B_2 units. The octahedra are further linked to each other perpendicular to this plane. How many electrons does M need to contribute for stability of the boron framework?

8.13. Several years ago the suggestion was made that hydrogen at very high pressures would not only be metallic but would also be a superconductor. More recently numerical calculations have indicated a superconductor with a critical temperature of around 230°K at about 400GPa. **8.16** shows the proposed structure for such a material. (a) Construct a molecular orbital diagram for the H_{12} unit ($H(3)$ - $H(14)$) of the fragment of this structure shown in **8.17**. The H-H distances within each sheet are 1.38Å and those between the sheets are 0.83Å. In free H_2 the H-H distance is 0.74Å and has a bonding orbital with an energy of -17.4eV and an antibonding orbital at 4.2eV. The energy of an

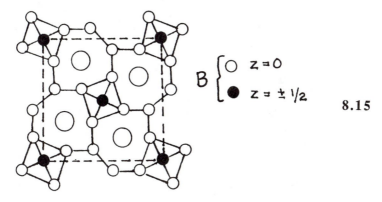

$$B \begin{cases} \bigcirc & z = 0 \\ \bullet & z = \pm \frac{1}{2} \end{cases}$$

8.15

isolated 1s orbital is -13.6eV. (b) Now allow the orbitals of the H_{12} unit to interact with those of the central H_2 unit (atoms $H(1), H(2)$). (c) In the 'real' structure for the superconductor, the local environment for $H(3)$ - $H(9)$, $H(4)$ -

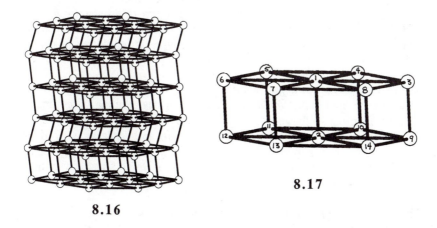

8.16

8.17

$H(10)$ etc., become equivalent to that shown for $H(1)$ - $H(2)$. Show how the *HOMO* and *LUMO* of (b) change when more pieces are added to the structure.

*8.14. Assemble the general form of the band structure of a tetrahedral solid, such as diamond, in the following way. (a) Combine s and p orbitals on the same atom to form sp^3 hybrid orbitals. (b) Link a pair of such hybrids together to give bonding and antibonding combinations. (c) The hybrids you drew in (a) are not orthogonal (see Question 2.3). Use this result to generate energy bands for the two basis orbitals generated in (b). Use your results to show how the band gap in the Group 14 elements decreases on moving down the group, the most dramatic drop occuring on going from carbon to silicon.

8.15. The structure of CaB_6 is as shown in **8.18**. As you can see it is derived from the $CsCl$ structure by replacing one of the ions by a B_6 octahedron, linked to six neighbors through the faces of the cube. Explain why this solid has this structure.

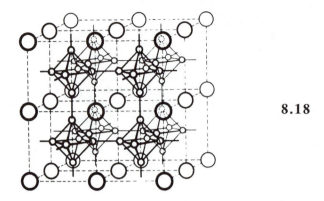

8.18

8.16. The band width of the 'e_g' band in $LaCoO_3$ with the perovskite structure appears to be considerably wider than the analogous band for $LiCoO_2$ with a derivative rocksalt structure. The CoO_2 part of the latter has the cadmium halide structure. Draw out the local geometry for a pair of adjacent Co atoms in each structure and by studying the overlap between the σ orbitals at one metal site with the metal orbitals at the next, provide an explanation for this difference.

8.17. $SnSe_2$ is a large band gap semiconductor with a resistivity of about 10Ωcm at room temperature. $CoCp_2$ is a molecular organometallic compound with a structure similar to that of ferrocene. However, when $CoCp_2$ comes into contact wth $SnSe_2$ it intercalates into the structure and a metal is produced which turns out to be a superconductor at temperatures less than about 6K. Explain why it becomes metallic. ($SnSe_2$ has a layer structure and the intercalation process involves the insertion of molecules between these layers.)

*8.18.The Scanning Tunnelling Microscope (STM) is able to probe the Fermi level of the surface density of states of metallic solids. Atoms where there is no such electron density are not imaged. In a (initially) very surprising result, only one half of the atoms of the graphite sheet may be imaged as shown in 8.19. This problem may be solved as follows. (a) Set up the Hamiltonian matrix for the energy bands associated with the $p\pi$ orbitals of a single graphite sheet and solve for the energy of the bands as a function of k. Since there are

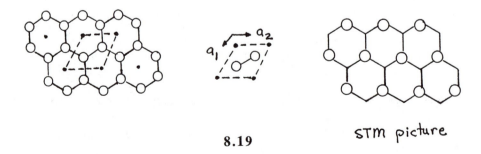

8.19   STM picture

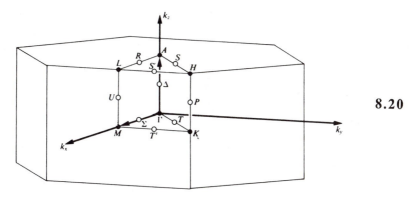

8.20

two $p\pi$ orbitals per cell, this will be a 2x2 determinant leading to two energy bands. Sketch out the band energetics connecting the points Γ, M and K of the

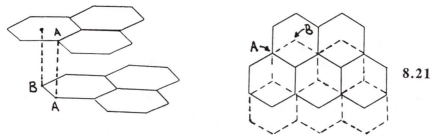

8.21

hexagonal Brillouin zone (**8.20**) as a function of the in-plane $p\pi$ interaction integral β. (b) Write down the form of the orbitals at the point K. Comment on the density of states at the Fermi level for the two carbon $p\pi$ orbitals of the unit cell. (c) **8.21** shows the three-dimensional structure of graphite. Notice that half of the atoms of the top sheet (labelled A) lie directly above atoms in the sheet beneath, whereas the other half (labelled B) do not. Although the inter-sheet distance is about 3.3Å the value of H_{ij} (call it β') linking two 'end-on' carbon $p\pi$ orbitals on the atoms A is about 0.6eV. Thus there may be significant dispersion perpendicular to the sheets. Calculate the dispersion along the line K (1/3,1/3,0) to H (1/3,1/3,1/2) using as a basis the two orbitals you derived in (b) for a unit cell containing two sheets. Hence show that the electron density around the Fermi level is much larger for the atoms of type A than of type B.

Answers

8.1. The $p\pi$ orbitals of an n-membered ring transform as each of the n irreducible representations of the cyclic group C_n. Because of time reversal symmetry the irreducible representations with complex characters fall into degenerate pairs as shown in the tables of Chapter 1 for these groups. As a result the energy levels for an N-membered ring are given within the Hückel approximation as $e_j = \alpha + 2\beta cos(2j\pi/N)$ and the molecular orbital coefficients as $N^{-1/2}\sum_{p=1,N}exp(2\pi ij(p-1)/N)\phi_p$. One graphical way of showing the

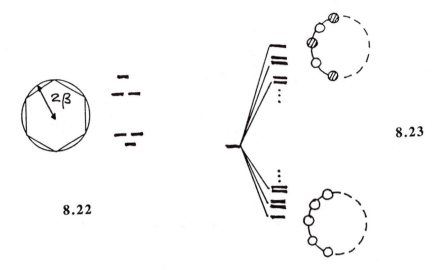

8.22

8.23

distribution of energy levels is to inscribe within a circle of radius 2β the polygon in question with one vertex at the very bottom (a Frost circle, **8.22**). The energy levels then occur where the vertices intersect the circle as shown for benzene. The highest energy orbital is antibonding between all pairs of adjacent carbon atoms, and the lowest energy orbital is bonding between all such pairs. The number of nodes in the wavefunction increases by one as the stack is climbed. (See Question 6.2 for a group theoretical derivation of the benzene energy levels.) Thus for a $2m + 2$ membered ring the energy levels occur in pairs except for the very lowest, and very highest ones as in **8.23**.

(a) Now for the infinite solid which may be mimicked as in **8.1** (the Born-von Kármán boundary conditions) there are an infinite set of levels, **8.24**. However from the Frost circle we know that this set is bounded at the bottom by one level with an energy of $e = \alpha + 2\beta$, and at the top by one level with an energy of $e = \alpha - 2\beta$. This collection of energy levels is called an energy band. It has a width of $/4\beta/$. The orbital description is very similar to that for the molecular case, the highest energy orbital antibonding between all (N) pairs of adjacent carbon atoms (N nodes), and the lowest energy orbital bonding between all such pairs (no nodes). In between are wavefunctions with $< N$ nodes. (b) The electronic configuration of polyacetylene is obtained by

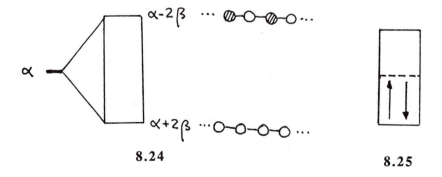

8.24

8.25

counting the number of electrons in the energy levels which make up the band. Since each carbon atom has one electron in a $p\pi$ orbital, overall the set of $p\pi$ orbitals will be half full, as in benzene, for example, where the three bonding orbitals are filled and the three antibonding ones empty. Thus the $p\pi$ band of polyacetylene *with this structure* will be half full and can be represented as in **8.25**. (c) Partially filled energy bands give rise to metals, systems which are electronic conductors. Thus if polyacetylene had this structure it would be metallic. In fact the structure of this material is a little different (see Question 8.5) and it is only a metallic conductor if some of the electrons in this filled band are removed, *i.e.*, by reaction with an oxidant (*e.g.*, Br_2) which removes a small number of electrons from the band.

8.2   (a) The square-planar $Pt(CN)_4{}^{2-}$ unit has the very characteristic d-orbital splitting pattern described in Question 3.11. Since the molecule contains Pt^{II}, the d electron count is eight and so the z^2 orbital, the *HOMO* of the unit, is full with two electrons. (b) In the solid the z^2 orbital is broadened into a band as shown in **8.26**. At the bottom of the band the basis orbitals are mixed in-phase, and at the top out-of phase as shown. There are two electrons in the z^2 band of the solid, sufficient to fill it. Recalling that full energy bands lead to insulating or semiconducting behavior, $K_2Pt(CN)_4$ is expected to be an insulator as it is. (c) Addition of bromine leads to oxidation of the system. Addition of δ atoms of bromine per formula unit thus leads to oxidation of the platinum atom so it has $8 - \delta$ electrons. The z^2 band now contains $2 - \delta$ electrons (**8.27**). (d) Since

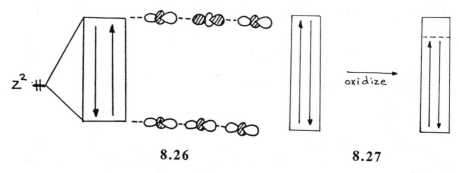

8.26   **8.27**

these δ electrons are removed from the top of the z^2 band where the wavefunction is maximally antibonding, the Pt-Pt bond is strengthened. This shows up in a contraction of the Pt-Pt distance from about 3.3Å to 2.9Å on bromine oxidation. In addition the material, with its partially filled band, becomes metallic. Crystals of the solid have a bronze appearance and are in fact as good a metal as graphite.

8.3.   We first have to consider two very different descriptions of electrons in molecules and solids which can be illustrated by reference to the electronic description of molecular hydrogen. We start off with the traditional molecular orbital approach (Mulliken-Hund = *MH*) to the chemical bonding problem in the H_2 molecule. Ignoring normalization and antisymmetrization, and using ϕ_1 and ϕ_2 as the two hydrogen 1s orbitals, there are two electrons located in the H-H bonding orbital $(\phi_1 + \phi_2)$ and we can write a singlet wavefunction in the following way

$$^1\Psi_{MH} = (\phi_1 + \phi_2)(1)(\phi_1 + \phi_2)(2)$$
$$= \phi_1(1)\phi_1(2) + \phi_2(1)\phi_2(2) + \phi_1(1)\phi_2(2) + \phi_2(1)\phi_1(2)$$

The one-electron energy can be readily evaluated by setting $<\phi_1/\mathcal{H}^{eff}/\phi_1> =$ $<\phi_2/\mathcal{H}^{eff}/\phi_2> = \alpha$ and $<\phi_1/\mathcal{H}^{eff}/\phi_2> = \beta$ using the Hückel model. It is just $2(\alpha + \beta)$ where $\alpha, \beta < 0$. However, one result of this electronic description is that the two electrons are allowed to simultaneously reside on one of the two atoms since terms such as $\phi_1(1)\phi_1(2)$ and $\phi_2(1)\phi_2(2)$ occur in the wavefunction. The simplest way to estimate the two-electron part of the energy is to multiply the probability of finding the two electrons on the same atom by U, the electrostatic interaction energy of two electrons located on the same atom. This probability is 1/2, since in $^1\Psi_{MH}$ each of the four terms is equally weighted. The total energy is then

$$e_{M\text{-}H} = 2(\alpha + \beta) + 1/2U$$

Now, as the *H-H* distance increases the chance of two electrons residing on the same atom becomes less likely and the molecular orbital model less appropriate. At infinite separation each neutral *H* atom holds just one electron and never two, in the lowest energy arrangement. To describe this state of affairs we would write a localized (Heitler-London = *HL*) wavefunction of the form.

$$^1\Psi_{HL} = \phi_1(1)\phi_2(2) + \phi_2(1)\phi_1(2)$$

The total energy of this arrangement is then

$$e_{H\text{-}L} = 2\alpha.$$

So in asking which arrangement will be more stable, the localized *H-L* state with an energy of 2α, or the delocalized *M-H* state with an energy of $2(\alpha + \beta) + 1/2U$, it is clear that the critical ratio is that of the two-electron to one-electron energy parameters U/β. When $|U/\beta|$ is large then the localized arrangement is appropriate, when it is small then the delocalized (molecular orbital) picture is a better description. In a solid a system described by the localised picture cannot be a metal since there is a Coulombic repulsion which prevents two electrons being on the same atom at the same time. Also the energy band approach of Question 8.1 is now inappropriate. Such a material is described as an antiferromagnetic insulator. Both pictures may coexist in the same material. For example consider a transition metal oxide. The interactions between the metal $(n+1)s$ and $(n+1)p$ orbitals and the oxygen $2s$ and $2p$ orbitals may well be strong enough (*i.e.*, large β) to lead to a *M-H* picture, but the corresponding interactions involving the nd orbitals may be much weaker and result in a *H-L* description. The answer to Question 4.31 shows how localized and delocalized pictures may coexist in molecules in Rydberg states. The balance between the one- and two-electron terms in the energy also controls the relative stability of high-spin (large $|K/\Delta|$) and low-spin (small $|K/\Delta|$) transition metal complexes. Here K is the exchange energy and Δ the e_g/t_{2g} splitting.

   With reference to the materials shown in (a) and (b) we need to investigate the dependence of the critical parameter $|U/\beta|$ on the nature of the transition metal. We know that as the size of the atom increases the magnitude of the electron-electron interactions, U and K (measured by the Racah parameters) decreases. This is shown in the size of the nephelauxetic effect in electronic spectra. In terms of the one-electron part of the energy we know too that the e_g/t_{2g} splitting (for example) in octahedral complexes of the transition metals is larger for second and third row metals than for first. This is clear from the colors of analogous complexes; first row ones frequently absorb in the visible (and are therefore colored) but their second and third row analogues absorb in the ultraviolet (and are therefore colorless). The metal $3d$ orbitals are quite contracted for the first row transition metals, so that for typical metal-ligand distances β will be smaller than for the heavier analogues. Associated with contracted orbitals are large values of U and K. For the larger second and third row metals with less contracted orbitals, β is larger and U and K smaller by comparison. Thus on both counts $|U/\beta|$ is expected to be larger for the nickel system of **8.4** than for the platinum analogue. Indeed octahedral complexes of Ni^{II} are often paramagnetic and octahedral, but those of Pt^{II} diamagnetic and square planar (see Question 3.11). In this light the

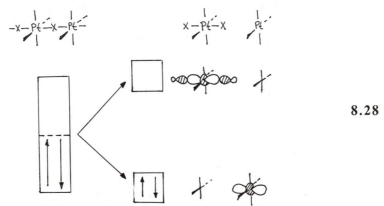

8.28

antiferromagnetic insulator found for the nickel system is not unexpected. Analogously the platinum system should be described by a delocalized or band model. Electron counting tells us that the platinum is present as Pt^{III} with a d^7 configuration. Thus the z^2 band, the highest occupied band of the solid, is half full. Such an arrangement is unstable to a Peierls distortion shown in **8.28**. The structure distorts to give a square planar, Pt^{II}, center and an octahedral Pt^{IV} center as may be readily seen from the orbital description of the bands.

8.4.   This is a simple problem since the three p orbitals on each center form three mutually orthogonal energy bands as shown in **8.29**. Also shown schematically, is an s band split off from them. This is a simplified, but useful, model for the simple cubic structure. For a Group 15 element (phosphorus, for example) the electron count (s^2p^3) is sufficient to completely fill the s band and

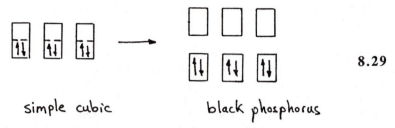

8.29

simple cubic   black phosphorus

half-fill the set of three *p* bands as shown. This is then the three-dimensional analogue of polyacetylene (Question 8.5) and we expect to see a Peierls distortion, with alternating short and long distances, along each of the three perpendicular x, y and z directions as in **8.30**. This is just what is found in

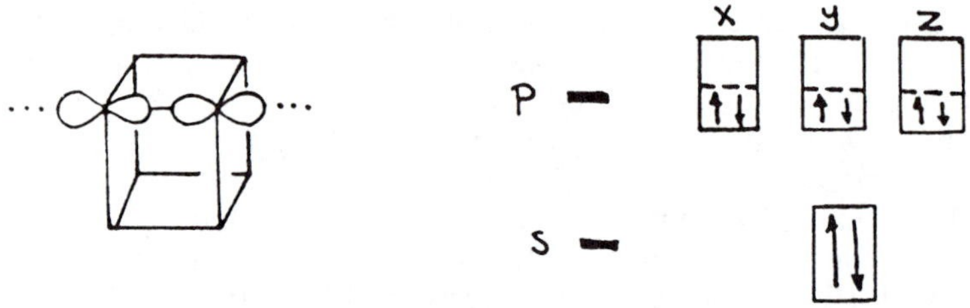

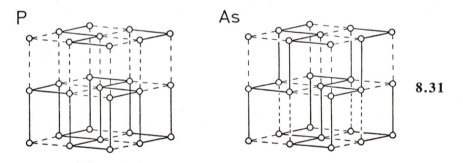

8.30

8.5, the structure of black phosphorus. A very similar structure is found for the structure of α-arsenic. They differ only in the way the three perpendicular ...short-long-short-long... structures are arranged as shown in **8.31**.

P   As

8.31

8.5.   There are two orbitals in the unit cell for such a problem, shown in **8.32**. The Bloch functions are thus:

$$\psi_1(k) = N^{-1/2}(...\phi_1 exp(-ika') + \phi_2 + \phi_3 exp(ika')...)$$

$$\psi_2(k) = N^{-1/2}(...\phi_4 exp(-ika'/2) + \phi_5 exp(ika'/2) + \phi_6 exp(3ika'/2)...)$$

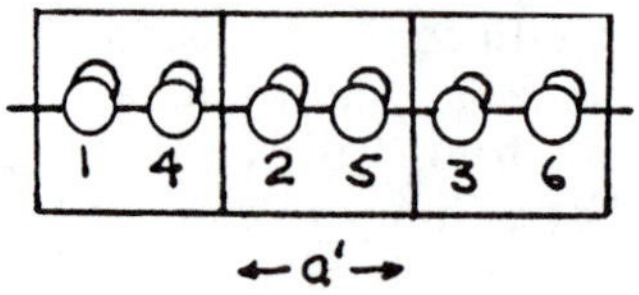

8.32

←a'→

To set up the secular determinant we need the H_{ij}'s. The diagonal elements are easy since there is no interaction between ϕ_1 (say for ψ_1) in one cell and ϕ_2 in the next. Thus :

$$H_{11} = <\psi_1(k)|\mathcal{H}^{eff}|\psi_1(k)> = N^{-1/2}N^{-1/2}(N\alpha) = \alpha,$$

$$H_{22} = <\psi_2(k)|\mathcal{H}^{eff}|\psi_2(k)> = N^{-1/2}N^{-1/2}(N\alpha) = \alpha.$$

and both are independent of k. The off-diagonal term does however, contain k. It is

$$H_{12} = <\psi_1(k)|\mathcal{H}^{eff}|\psi_2(k)> = N^{-1/2}N^{-1/2}(N(exp(ika'/2) + exp(ika'/2))\beta$$
$$= 2\beta cos(ka'/2)$$

The secular determinant is thus

$$\begin{vmatrix} \alpha - e & 2\beta cos(ka'/2) \\ 2\beta cos(ka'/2) & \alpha - e \end{vmatrix} = 0$$

with roots $e = \alpha \pm 2\beta cos(ka'/2)$. This situation is sketched out as in **8.33**.

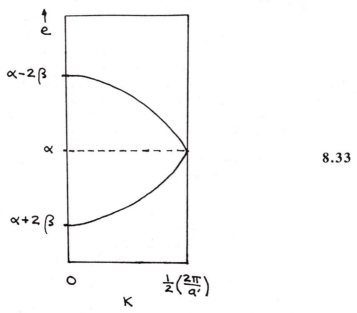

8.33

The situation shown in **8.34**, where the bond lengths alternate, differs in the value of H_{12} which enters the secular determinant. now it is given by

$$H_{12} = <\psi_1(k)|\mathcal{H}^{eff}|\psi_2(k)> = N^{-1/2}N^{-1/2}(N(\beta_1 exp(ikxa') + \beta_2 exp(ik(1-x)a'))$$
$$= \beta_1 exp(ikxa') + \beta_2 exp(ik(1-x)a'$$

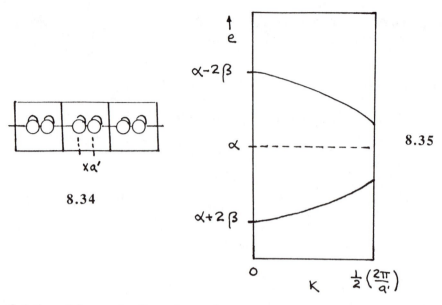

8.35

8.34

Solution of the secular determinant gives

$$e = \alpha \pm (\beta_1{}^2 + \beta_2{}^2 + 2\beta_1\beta_2 cos(ka'))^{1/2}.$$

This state of affairs is shown in **8.35** and shows two energy bands with a gap at the zone edge. The energetic preferences of the half-filled band situation are interesting. The energy of the bottom band at $k = 0$ is just $e(0) = \alpha + (\beta_1 + \beta_2)$, and at the zone edge $e(1/2) = \alpha + (\beta_1 - \beta_2)$. Compare these results with the comparable values of $e(0) = \alpha + 2\beta$, and $e(1/2) = \alpha$ for the equidistant chain. If $\beta \sim (\beta_1 + \beta_2)/2$, then it is the electrons in levels close to the zone edge which are stabilized on distortion. Schematically, using boxes for bands the energetic situation is described as in **8.36**. Polyacetylene with one electron per π orbital, and thus a half-filled band, will distort away from the situation with equal *C-C* distances and an insulator or semiconductor will result.

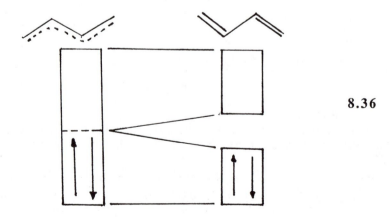

8.36

8.6. The $/H_{ii}/$ values of the oxygen and titanium orbitals decrease in the order $O(2s) > O(2p) > Ti(3d) > Ti(4s) > Ti(4p)$. Thus we expect five energy bands largely associated with these orbitals in this order. The d band will be split into two as a result of the octahedral environment of the metal (e_g and t_{2g} bands), and there will be some splitting in the $O(2p)$ band (separating σ and π type interactions) as shown in **8.37** and **8.38**. The largely metal orbitals are all antibonding, wheareas the orbitals which are largely oxygen in character are bonding.

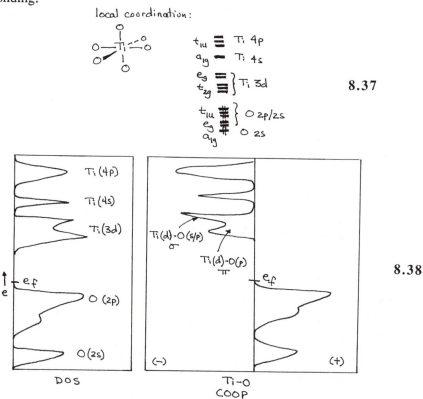

8.7. (a) The molecular orbital diagram for a square planar SnH_4 molecule is assembled in **8.39**. The a_{2u} and b_{1g} orbitals lie close in energy and could be reversed in energy.

(b) The dispersion of a given energy band (its band width) depends on the size of the overlap from cell to cell. For orbitals which are symmetric with respect to reflection in the molecular plane (e.g., the e_u and a_{1g} orbitals) the most bonding combination will lie at $k = 0$ where they are in-phase. For orbitals which are antisymmetric with respect to reflection in the molecular plane (e.g., the a_{2u}) the most bonding combination will lie at $k = (1/2)(2\pi/a)$ where they are in-phase. Using these two considerations the band structure may be drawn as in **8.40**. The largest dispersion is drawn for the a_{2u} orbital (σ overlap), smaller dispersion for the e_u orbitals (π overlap) and a_{1g} orbitals (σ overlap but less directional than that for a_{2u}) and zero dispersion for the b_{1g} orbital

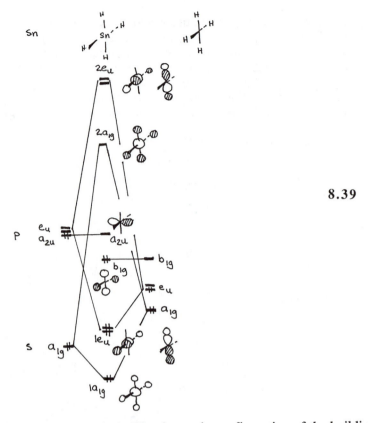

8.39

(no central atom overlap). The electronic configuration of the building block is sufficient to fill the $1a_{1g}$ and $1e_u$ bands (six electrons) but the remaining two must half fill the a_{2u} and b_{1g} bands. At this geometry the a_{2u} band will be half full of spin-paired electrons (wide band) but the b_{1g} band full of spin parallel electrons. Such an electronic arrangement is susceptible to a Peierls distortion in which $Sn\text{-}Sn$ distances alternate (see Question 8.5).

8.8. (a) *Sr* is very electropositive, thus, we can consider the solid to be effectively $Sr_2^{2+}[TiO_4^{4-}]$. Each *Ti* atom is octahedrally coordinated so the density of states (*DOS*), ignoring the *Sr*-based orbitals is readily derived as in **8.41**. Notice from the *COOP* curve the bands from -16.7eV to -14.5 (largely oxygen located) are $Ti\text{-}O$ bonding, while those from -11.0 to -9.0 (largely metal located) are antibonding.

    (b) There are three bands from -11.0 to -9.0eV. This is consistent with these bands being the "t_{2g}" members of each octahedron where the oxygen atoms are π antibonding to the metal. There are twelve bands in the -14.5 to -16.7eV region and they are $Ti\text{-}O$ bonding. Thus, they must be oxygen p (3 A.O.'s on each oxygen and 4 oxygens per unit cell = 12 bands) bonding to *Nb* $4d, 5s, 5p$. To figure the position of the Fermi level, we need to count the electrons in the unit cell. From $Sr_2^{2+}TiO_4^{4-}$ with O^{2-} we have Ti^{4+}, d^0. The Fermi level must then lie at the top of the oxygen p bands. (In fact, by

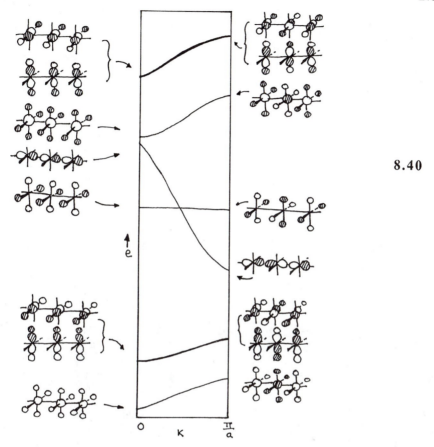

8.40

definition, the Fermi level in insulators lies halfway between the occupied and unoccupied bands. Only in metals is the Fermi level that of the highest occupied level.)

(c) Since the bands in this region are metal-based members of the t_{2g} set we expect that they are antibonding to oxygen p in a π fashion.

The orbitals are shown in **8.42** starting at Γ, with a top view that shows only one out-of-plane oxygen (the other interacts in an equivalent fashion). A general feature is that the xy band can never interact by symmetry with the out-of-plane oxygen π orbital. Likewise, the xz and yz bands always interact with the out-of-plane π orbitals.

At Γ the xy band is lower than xz,yz because of the reason just given. On going from Γ to X both xy and xz are destabilized since oxygen p A.O.'s enter in an antibonding fashion in the x direction whereas yz stays at constant energy. On going from X to M both yz and xy gain another antibonding interaction so both rise in energy. The xz orbital has the same form along this line so it stays at constant energy. Notice that at Γ and M xz and yz are degenerate, by symmetry.

8.9. (a) Since oxygen is much more electronegative than lead we can write the solid as $Pb^{2+}O^{2-}$ so that Pb has 2 electrons and oxygen has 8. The basic

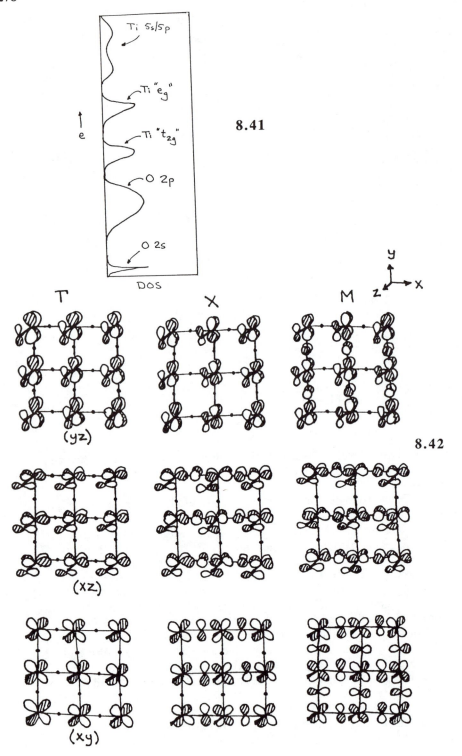

8.41

8.42

structural entity is thus **8.43**. Let us forget about the lone pairs around oxygen and consider only the σ electrons. At each oxygen there is a σ type hybrid so the total electron count at each tetragonal pyramid is $4x2 = 8$ electrons from this source plus the two on Pb^{2+} to give a total of 10 electrons. For an AH_4 system where H is much more electronegative than A the level ordering is shown in **8.44**. The highest occupied bands then will be derived from the $2a_1$ orbital. There will be two of them since there are two Pb atoms per unit cell. Their basic shapes are shown in **8.45**. They are lone pair orbitals on lead.

8.43

8.44

8.45

Notice the six bands labeled $2e$, $2b_2$, $1e$ and $2a_1$ are derived from the oxygen p A.O.'s. There are two at lower energy, not shown in **8.12**, that come from oxygen $2s$ orbitals.

(b) When the PbO layers are stacked on top of each other, not too much will happen to the six oxygen p based bands since there is no sheet-to-sheet O-O overlap. They have approximately the same dispersion as in the two dimensional (1 layer) band structure calculation in **8.12**. It is clear from the *COOP* curve that they must be in the -14.5 to -17.5 eV region. There should be Pb-O bonding in this region (see the $1e$ set for AH_4). There is a smaller *DOS* in the middle of this region because there is only one band, labeled $2b_2$, which cuts across this energy range. However, since "$3a_1$" and "$4a_1$" are directly pointed towards adjacent layers, they will become much more dispersive on moving from two to three dimensions. The region from -13.5eV to -10.5eV contains Pb-O antibonding character. This requires that the $2a_1$ derived bands in AH_4 have oxygen p character as well. Thus the $3a_1$ and $4a_1$ orbitals are more like those shown in **8.46**. Notice that while $3a_1$ is Pbs -O s antibonding, it is Pb p-O s bonding. The interlayer Pb-Pb interaction is mainly derived

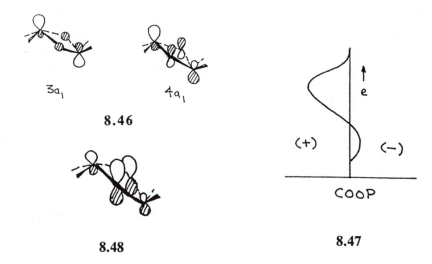

3a₁   4a₁

8.46

e

(+)   (−)

COOP

8.48   **8.47**

from *3a₁* and *4a₁*. One would expect a *COOP* curve like that shown in **8.47** and that is basically what is found. The bottom is *Pb-Pb* bonding, the top antibonding and the middle nonbonding. Notice that there is also a small *Pb-Pb* bonding peak at -17eV. This is due to the bonding equivalent of *4a₁* which contains some *Pb* character (**8.48**).

(c) The *3a₁* and *4a₁* bands show dispersion in 3 dimensions when the layers are stacked. Since both bands are filled this should be a repulsive

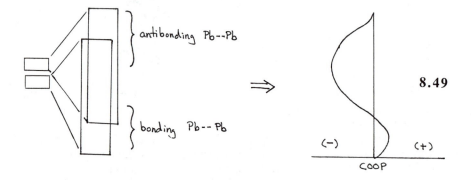

antibonding Pb--Pb

⇒

8.49

bonding Pb--Pb

(−)   (+)

COOP

interaction when overlap is included as in **8.49**. As shown in this *COOP* curve the antibonding peak should be larger than the bonding one. Yet, from the real *COOP* curve in **8.13**, there is less antibonding than bonding. This suggests that higher lying orbitals have mixed into the top of the *3a₁* and *4a₁* derived bands in a way to reduce the *Pb-Pb* antibonding. If you look at the α *PbO* structure the *Pb* atoms are not located directly over each other but in a staggered arrangement. The "*AH₄*-like" empty orbitals are *Pb*-centered and correspond to *3a₁* and *2e* of **8.44**. A net attractive interaction between the *PbO* layers results as in **8.50**. This is a similar effect to that described in Question 5.11.

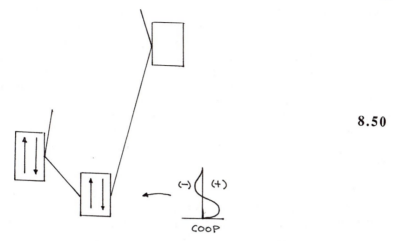

8.50

COOP

8.10. The sphalerite structure is one where all of the carbon atoms of diamond have been substituted alternately by zinc and sulfur. This is a stable structure for an average of four electrons per atom. **8.51** shows a schematic band structure for the solid. (It is derived in more detail in Question 8.15.) With more than four electrons per center, electrons have to be placed in antibonding orbitals, some of the bonds are broken and the structure is destroyed. Thus GaS (4.5 electrons per atom) has a structure where some of the atoms are three

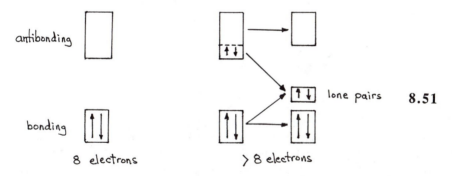

8.51

coordinate. The extra electrons form lone pairs at these three-coordinate sites. This is shown schematically in **8.51**. Thus each atom still obeys the octet rule. In $CdIn_2Se_4$ where there is a total of $(2+(2x3)+(4x6)=)$ 32 electrons (or 4.57 electrons per atom), the structure is still intact but some of the tetrahedral sites in the crystal are unoccupied. Lone pairs of electrons point into such sites. This is another way to stabilize a structure with more than four electrons. Simple electron counting allows us to generate a simple rule (the Grimm-Sommerfeld rule) to describe this. Taking into account the electronic demands of these three and four-coordinated atoms, in all tetrahedrally-based solids there should be an average of four electrons per site, irrespective of whether the site is occupied or not ($\square$). Thus in $\square CdIn_2Se_4$ there are now 32/8 = 4 electrons per atom to be compared with 32/7 = 4.57 electrons per atom. We could visualize the formation of the defect structure by analogy with those of the boranes by using the following Gedanken, or thought, process. A given electron count is associated with stability of a given structure. This is four electrons per atom for the octet solid and $(2n + 2)/n$ electrons per site for an n-vertex *closo* deltahedron. Addition of extra electrons leads to instability. One way of relieving this is to eject an atom from the structure. In the sphalerite case this gives rise to a defect structure, and in the borane case to a *nido* deltahedron. In both cases the number of electrons per site (including the vacancy) is the same as in the parent. Recall that the *nido* deltahedron has the same number of skeletal electrons as the *closo* form. In $CuIn_2\square_xSe_3Br$, there are a total of $(1+(2x3)+(3x6)+(1x7)/(7+x)$ electrons per site. To satisfy the Grimm-Sommerfeld rule $x = 1$.

8.11. **8.52** shows the derivation of the band structure of the two-dimensional CuO_2 sheet. We start off with the orbitals of an axially distorted octahedron. As is clear to see the highest occupied band is derived from the $x^2 - y^2$ orbital

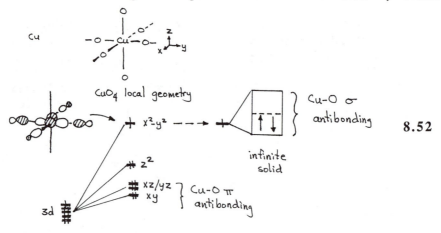

of the metal, which is certainly metal-oxygen antibonding. As electrons are withdrawn from this band, by replacing of some of the La^{3+} atoms with Sr^{2+} then the antibonding effect is reduced and the bond lengths are expected to shrink, as observed experimentally. This is in spite of the fact that the Sr^{2+} ion is larger (*i.e.*, has a larger ionic radius) than the La^{3+} ion.

8.12. Consider an M_2B_8 unit which contains a B_6 octahedron and an ethylene-like B_2 unit. The B_6 octahedron requires 6 electrons for extra-octahedral bonding plus 14 electrons for skeletal bonding. The ethylene unit requires 4 electrons to bond to its neighbors and 4 electrons ($2\sigma + 2\pi$) internally. The total number of electrons need is 28 in total. 8 boron atoms provide 24 and so each metal must provide 2.

8.13. (a) There are a number of ways to determine the orbitals of the H_{12} fragment. Perhaps the easiest is to take 6 H_2 σ and 6 H_2 σ^* type orbitals and split them according to the D_{6h} symmetry in H_{12} in Question 2.4 as shown in **8.53**. Notice that the symmetry species of the σ and σ^* sets are all different. Therefore, there is no intermixing between them. The placement of the levels simply follows the *H-H* distances. The spread of the a_{1g} to b_{1u} orbitals will be smaller than that for a_{2u} to b_{2g} since the orbital coefficients in the σ^* set are larger and hence lead to a larger overlap. This same result of course could have been obtained by a "brute force" approach with group theory using the *s* orbitals of *H(1)* to *H(12)* as a basis.

(b) Only two orbitals of H_{12} are involved as shown in **8.54**. All other orbitals are exactly as drawn in **8.53**. Notice that $2a_{1g}$ and $2a_{2u}$ rise to very high energy. This is because in the H_{12} unit each H_2 segment overlaps with only the two adjacent H_2 units on either side to any appreciable extent. The H_2 unit in the middle, however, overlaps with all *six* H_2 units around it.

(c) As additional H_2 units are added the $2a_{1g}$ orbital gets more and more destabilized -- it will become maximally antibonding within each hexagonal layer and bonding between the two layers. Meanwhile $1a_{2u}$ becomes totally bonding within each layer and antibonding between them to generate the energy 'bands' of **8.55**.

8.14. The generation of a qualitative band structure is shown in **8.56**. Shown are the construction of the sp^3 hybrids with an energy of $e_{hy} = 1/4(e_s + 3e_p)$, their interaction to give a bonding and antibonding pair separated in energy by Δ, and finally, by allowing each of these combinations to interact with similar hybrids on the same atom with an energy of $1/4(e_s - e_p)$, the formation of two energy bands. By analogy with the answer to Question 8.1, the width of each of

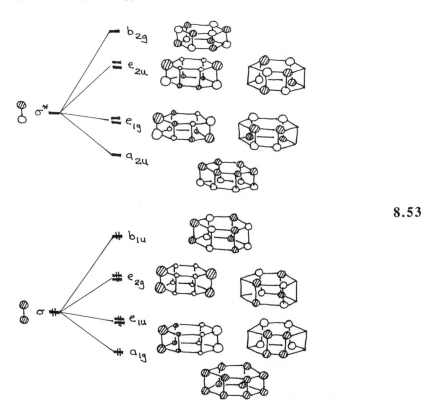

8.53

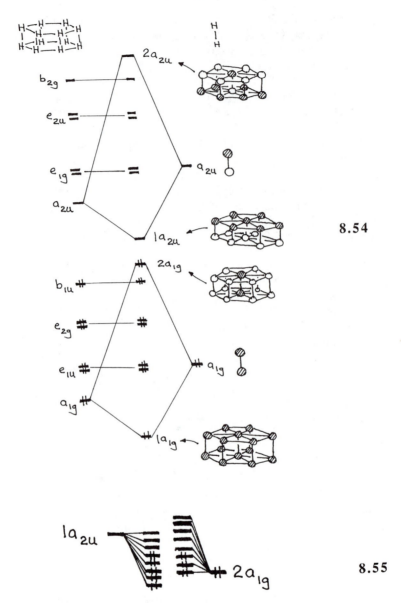

8.54

8.55

the bands is approximately $|(e_s - e_p)|$. The band gap, E_g, is then roughly equal to $\Delta - |(e_s - e_p)|$. The term $(e_s - e_p)$ does not change dramatically on moving down the group, but Δ does. Roughly speaking Δ varies inversely as the square of the internuclear distance. The largest change (Table 8.1) is from carbon to silicon.

An interesting result comes from comparison of your answers to Questions 8.1, 8.5 and 8.15. It should be clear that one of the criteria for the generation of an insulator for the half-filled band situation is that there are two different interaction integrals between the orbitals of the problem. In the case of the

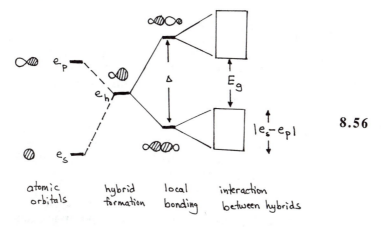

| atomic orbitals | hybrid formation | local bonding | interaction between hybrids |

8.56

Table 8.1 Single bond distances in the Group 14 elements

Element	Single Bond Distance, d (Å)	$1/d^2$
carbon	1.54	1.00[a]
silicon	2.34	0.43
germanium	2.44	0.40
tin	2.80	0.30
lead	2.88	0.29

a. Scaled relative to the value for carbon

alternating chain these are β_1 and β_2, and in the diamond case, the different inter- and intra-atom interaction integrals between the atomic hybrids.

8.15. Each B_6 unit requires 14 electrons for skeletal bonding and six electrons which will be involved in forming two center-two electron bonds to other octahedra through the faces of the cube formed by the Ca ions. This total of 20 electrons is provided by six boron atoms (18 electrons) and one Ca atom (2 electrons). In fact both KB_6 and LaB_6 are known with this structure, respectively with one less and one more electron than CaB_6.

8.16. **8.57** shows the local geometries in the two structures. An important point to note is that for the cadmium halide structure the metals are connected

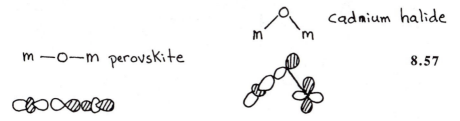

cadmium halide

m —o—m perovskite

8.57

to each other *via* an oxygen atom and an angle of 90°, but in the perovskite structure this angle is 180°. Thus the metal σ orbitals at one center are connected to those of the next metal atom by a π interaction in the case of $LiCoO_2$ but a σ interaction in $LaCoO_3$. The $d\sigma$ -O-$d\sigma$ overlap in the perovskite structure gives rise to a larger bandwith than the $d\sigma$-O-$d\pi$ overlap in the cadmium halide structure.

8.17. $CoCp_2$ is a nineteen electron organometallic compound, and as such is easily oxidized. Intercalation into $SnSe_2$ leads to population of the conduction band of the solid leading to a metal- a characteristic of a partially filled band. As yet the chemical requirements for a superconductor are not known.

8.18. (a) Since there is no nearest-neighbor overlap between the orbitals of type A (or B) in one cell, and those of the same type in adjacent cells the diagonal entries of the secular determinant are simply α - e. The off-diagonal entries are simply the sum of the $\beta exp(ik.r)$ terms where r is the vector linking pairs of nearest-neighbor atoms. This is readily evaluated as H_{12} = $\beta(exp(ik.(2/3a_1 + 1/3a_2)) + exp(ik.(1/3a_1 - 2/3a_2)) + exp(ik.(1/3a_1 +1/3a_2)))$. Thus the energy of the orbitals as a function of k is given by the roots of the determinant

$$\begin{vmatrix} \alpha - e & H_{12} \\ \\ H_{12} & \alpha - e \end{vmatrix} = 0$$

which are $e = \alpha \pm A^{1/2}\beta$ where $A = [3 + 2cos(a_1 + a_2).k + 2cos(a_1.k) + 2cos(a_2.k)]$. Substitution for values of k lead to energies of $e(\Gamma) = \alpha \pm 3\beta$, $e(M) = \alpha \pm \beta$ and $e(K) = \alpha \pm 0\beta$. A 'spaghetti' diagram is shown in **8.58**. Since each $p\pi$ orbital contains one electron these bands are half full and the Fermi level lies at the K point. (b) This point where the two bands both have energies of α, the free $p\pi$ value, are therefore nonbonding. There are of course an infinite number of different ways to write the wavefunctions for this

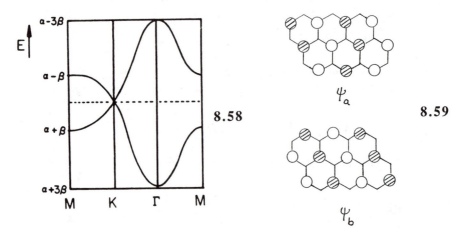

8.58

8.59

degenerate pair (just as for the nonbonding pair of orbitals in cyclobutadiene in Question 6.7) but one pair which will be useful in (c) is shown in **8.59**. They show that the electron density at the Fermi level will be identical for the two carbon atoms. Indeed all properties should be the same for the two- they are symmetry equivalent. (c) The dispersion along a_3 for the $p\pi$ orbitals of type A is clearly zero (**8.60**) since there is no interaction between orbitals of this type on adjacent sheets. The situation though is very different for the orbitals of type B. Here there is a nonzero interaction integral between them. In this direction there will be a very simple energetic dependence on k. This will just be a one-dimensional band of the type described in Questions 8.1 and 8.5 where there are two orbitals per unit cell (one on each sheet). Using the result

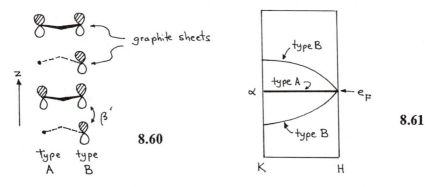

8.60   8.61

derived in the answer to Question 8.5 the dispersion perpendicular to the graphite planes for these atoms will be given by $e = \alpha \pm 2\beta' \cos(kc/2)$. The dispersion curves for the two types of $p\pi$ orbital are shown in **8.61**. The diagram shows very clearly that the larger electron density at the Fermi level will be associated with the type A atoms, and that associated with the type B atoms will lie deeper in energy. Perhaps the most suprising result here is that the interaction energy is large enough to make a difference. By calculation $\beta' \approx 0.6\text{eV}$ a value consistent with the energy window around the Fermi level that may be sampled experimentally.